Jonathon _____ eminent writer, broadcaster and campaigner on sustainable development. Established in 1996, Forum for the Future is now the UK's leading sustainable development charity, with 70 staff and over 100 partner organisations, including some of the world's leading companies

In addition, Jonathon is President of Population Matters, President of The Conservation Volunteers and a Director of Collectively (an online platform celebrating sustainable innovation). He was formerly Director of Friends of the Earth (1984–90), co-chair of the Green Party (1980-83), of which he is still a member, a Trustee of World Wildlife Fund UK (1991–2005) and between 2000–2009 he was Chair of the UK Sustainable Development Commission, providing high-level advice to government ministers.

Jonathon was installed as the Chancellor of Keele University in February 2012 and he received a CBE in January 2000 for services to environmental protection.

Praise for *Hope in Hell*

'Yet another brilliant book from Jonathon Porritt. An eloquent, thoughtful and compelling affirmation that we can address the climate crisis if we set our minds to it.'
Christiana Figueres

'*Hope in Hell* provides a brilliant analysis of humanity's impact on the Earth. Jonathon Porritt still allows us a little hope, but absolutely no excuses for further delay, urging radical political action. Don't read this unless you are prepared to strive for the rest of your days in service of future generations; you will be emotionally enlisted, and unable to claim ignorance in your defence.'
Helen Browning, Soil Association

'It's a terrific book and couldn't be more timely – the post-COVID world will need just this kind of synoptic picture and I hope the book will be recognised as a major tract for the times.'
Rowan Williams

'*Hope in Hell* will become an indispensable handbook in the pre-eminent planetary struggle of our times. It is truthful, trenchant and yet refreshingly hopeful.'
John Vidal

'*Hope in Hell* provides a blueprint of how we can rapidly decrease emissions, lessen inequalities and capitalise on this unique moment for change, ensuring a better world for all. Essential reading.'
Rosie Boycott

'Jonathon Porritt draws on five decades of experience to present this vital book, one that will change how you think about climate change and transform what you will want to do about it – hopefully, just in time to save our collective future.'
Tony Juniper, environmentalist

'Despite daunting obstacles, and a rapidly shrinking window of opportunity, Jonathon Porritt argues that the balance of factors at play can still lead to a positive outcome as we grapple to find our balance within the natural world. Will we make it? Not without confronting the status quo and the elites that defend it, with civil disobedience and the solidarity of the engaged young and old offered as crucial ingredients. The decade of the 2020s will be decisive, and *Hope in Hell* offers a blueprint for determining which fork in the road we will take.'
Chris Rapley CBE, Professor of Climate Science at University College London

'This book offers real hope as to how we might re-set our economies, post the coronavirus crisis. But that hope has to be earned; as Porritt puts it: "There is no hope whatsoever in another ten years of incremental change." Radical transformation is needed, including mass civil disobedience – this really is the last chance saloon for avoiding climate-driven societal collapse. When Jonathon Porritt says this, the world has to listen.'

Prof. Rupert Read, University of East Anglia, author of *This Civilisation is Finished* and national spokesperson for Extinction Rebellion

'*Hope in Hell* sits at a pivotal moment in the history of humanity and the planet. It rehearses our most dread fears and rigorously explores the terrors of what is likely to come. But it is threaded through with just enough hope to promise a very different future for our children – if, that is, we take the right actions, right now. This book mixes intellectual acuity, fastidious research, a call to arms and a deeply personal plea. That plea has touched and inspired me. We should all read this important book.'

Kevin McCloud

'An essential resource if you have any hope for the unknown future of yourself, your children and all life on this planet.'

Mark Rylance

'Not for the first time, Jonathon Porritt has put his finger on the pulse . . . the racing pulse of the whole of humanity in extreme danger as a consequence of the existential threat of climate change.'

David Puttnam

HOPE
IN HELL

A DECADE TO CONFRONT
THE CLIMATE EMERGENCY

JONATHON PORRITT

**SIMON &
SCHUSTER**

London · New York · Sydney · Toronto · New Delhi

First published in Great Britain by Simon & Schuster UK Ltd, 2020
This edition published in Great Britain by Simon & Schuster UK Ltd, 2021

1 3 5 7 9 10 8 6 4 2

Simon & Schuster UK Ltd
1st Floor
222 Gray's Inn Road
London WC1X 8HB

www.simonandschuster.co.uk
www.simonandschuster.com.au
www.simonandschuster.co.in

Simon & Schuster Australia, Sydney
Simon & Schuster India, New Delhi

A CIP catalogue record for this book
is available from the British Library

Paperback ISBN: 978-1-4711-9330-9
eBook ISBN: 978-1-4711-9329-3

Typeset in Bembo by M Rules
Printed and bound by CPI Group (UK) Ltd, Croydon, CR0 4YY

MIX
Paper from
responsible sources
FSC® C020471

For all those
ready to embrace more radical responses
to today's Climate Emergency.

For young people today
already stepping up
with the kind of conviction, courage and compassion
on which our future now depends.

And for all the rest of us,
who now know where our duty lies.

Contents

INTRODUCTION

Few knew what lay ahead of us when the World Health Organization declared COVID-19 to be a global pandemic on 11 March 2020. This immediately prompted predictions that our time-bound world, from that point on, would be starkly divided into pre-COVID-19 times and post-COVID-19 times. Even now, it's not easy to wrap our minds around the shape of that post-COVID world, but we know for sure that we will *never* be going back to life as it was before March 2020.

Even now, the full cost of the pandemic remains incalculable, measured both through the emotional burden of millions of lives lost and through the economic burden of doing what we had to do to stop those human costs spiralling out of control. As of April 2021, the International Monetary Fund calculated the hit to global economic output at $28 trillion.

For wholly understandable reasons, with so many people's lives so painfully disrupted in so many different ways, it proved difficult to keep any kind of focus on the Climate Emergency through 2020 – even though it is now much more clearly recognised that accelerating climate change threatens everything we hold dear in terms of our hopes for a better world.

What we're witnessing is a particularly telling example of 'the tragedy of the horizon', with COVID-19 posing an immediate and unavoidable threat, with the lives of so many at risk, necessitating comprehensive, 'whatever-it-takes' interventions from government. By contrast, the Climate Emergency is still seen by most people today as a challenge for tomorrow. Even though our continuing failure to get to grips with it today, *right now*, is putting at risk the lives of countless millions of people in the future.

As we look forward to a world in which we've learned how to live with COVID-19, it's even more important to understand that the Climate Emergency poses an infinitely graver risk to humankind than COVID-19 – despite the fact that it has warranted so little political engagement over the years. That's the tragedy of the horizon: today *always* trumps tomorrow.

Unless, that is, the sheer, gut-wrenching trauma of COVID-19 causes us all to start thinking very differently about the future. At the very least, people have already begun to understand that COVID-19 is almost certainly just the first in a new wave of pandemics – caused in large part by our seemingly insatiable desire to go on abusing the natural world and its wild creatures, with no thought for the consequences to ourselves.

Experts have been warning for many years that most of the new diseases that have emerged since 1960 come from wild animals; the risk of pathogens jumping from animals to humans has always been there, but our constant encroachment on the world's rainforests and other habitats has multiplied those risks many times over. As have the global trade in wild animals and wild animal markets.

Governments could have put a halt to all those things, in part to reduce the risk of future pandemics. But they chose not to – with too many short-term fires to be fought, so many other priorities to be funded – and we (as voters) lacked the fire power to make it a 'must do'.

Will it be any different with the Climate Emergency, with politicians continuing to focus on what's easy, forever deferring the more difficult decisions?

We start from a position of weakness. Ever since the 1980s, the idea of governments acting decisively to protect the interests of their citizens, in all circumstances, has been steadily devalued. Ideologically, the balance between 'the role of government acting on behalf of all its citizens' and 'the role of the market acting on behalf of all consumers' has long since been lost. Like it or not, the so-called 'laws of the market' have for so long ridden roughshod over the 'common wealth' of citizens, communities and the environment that we have forgotten what that balance once looked like.

This is why what is happening in the USA is so crucial. When Joe Biden was inaugurated on 20 January 2021 as the 46th President of the United States, more than 4,000 people a day were dying as victims of COVID-19. With nearly 500,000 deaths in all, there was no national vaccination programme in place. Trump-supporting Republican politicians were continuing to make light of the pandemic, refusing to endorse basic public health measures, and vociferously calling for a 'return to pre-COVID times', whatever the consequences.

Within days of getting his feet under the presidential desk in the Oval Office, President Biden had set in train a quite remarkable, centralised vaccination programme. At the time of going to press, 45 per cent of US citizens have been vaccinated, with the expectation of reaching 75 per cent by August.

This matters in itself – in terms of lives saved, hopes and aspirations reborn. Set alongside the American Rescue Plan Act, passed back in March 2021, it tells us something very profound about what is happening in the USA. Biden's $1.9 trillion stimulus package is the closest thing to an old-fashioned Keynesian investment programme that the world has seen for a very long time – not just in terms of the direct payments made to US citizens, beefed up unemployment insurance and new child-care benefits, but through significant investments in public services, including $30 billion devoted to public transport, $350 billion for state and local governments to repair their broken-down infrastructure assets (water, sewerage, broadband and so on), and a further $30 billion for tribal governments and indigenous communities.

Hot on the heels of that package comes a new American Jobs Plan, promising an additional $2 trillion for further investment in physical infrastructure (including EVs, charging networks, grid upgrades and so on), as well as a massive infusion of support for in-home care workers looking after the old and disabled. This 'once-in-a-generation investment in America' will be paid for, in part, by reversing the massive tax breaks for companies introduced by Donald Trump.

The truth of it is that after more than four decades of stagnating wages and massive under-investment in the 'common wealth' of the United States, their government is ready to take back control of managing the economy from market zealots and central bankers. Contrast Joe Biden's words ('The Government isn't some foreign force in a distant capital. It's us. All of us. We the People') with that now infamous quip from President Ronald Reagan in 1986: 'The most terrifying words in the English language: "I'm from the Government, and I'm here to help."'

This seismic shift in redefining the role of government has huge implications for addressing the Climate Emergency. Throughout *Hope in Hell,* I've set out to explain why this is *the* decisive decade; if we do what we need to do by the end of the decade to avoid runaway climate change, however 'unthinkable' that may be to most politicians at the moment, then we'll have a fighting chance of ensuring a better world for humankind in the future. But if we fail to grip that challenge, then it's more than likely that today's young people will be looking back on COVID-19 as a relatively insignificant, short-lived perturbation.

In the Introduction to the first edition of *Hope in Hell,* written this time in 2020, I suggested there were good reasons to hope that the pandemic, in all its shocking intensity, make it more likely that we would indeed rise to that challenge:

- Just so long as we realise that the warnings from experts (on pandemics or climate change) must now inform *all* future policy, and that those who dismiss that expertise as 'fake news' are dangerous enemies of their own people;
- Just so long as we come to recognise ourselves once again as creatures of the Earth, governed by the laws of physics and the biological interdependencies of *all* living creatures;
- Just so long as we use this unprecedented shock to our

     way of life to rethink our basic values and, indeed our
     ultimate purpose as human beings;
- Just so long as the post-COVID recovery process puts
  the Climate Emergency at its heart.

One year on, it's all too clear that governments around the world have not yet gripped that opportunity. In March 2021, a comprehensive report from Oxford University's Economic Recovery Project (working in partnership with the IMF and the United Nations Environment Programme), looking back over the whole of 2020, revealed that only $1 in every $40 of the $14.6 trillion committed by the world's fifty richest economies had any 'positive green characteristics'. To be fair, most of that expenditure was for short-term measures (furlough schemes, loans and grants to businesses, and so on) but even if we look at the $2 trillion dedicated to rebuilding longer term economic prosperity, only 18 per cent can be said to have positive green characteristics – including all low-carbon and conservation-related commitments.

Worse yet, even on something as obvious as 'stop funding fossil fuels, and start funding renewables', we're still way off the pace. According to Energy Policy Tracker, governments punted $202 billion into fossil fuel businesses through 2020, with $183 billion going to renewables.

That's the world we still live in. One foot in the past, drenched in oil, stamping heedlessly on every part of the natural world. One foot in the future, learning how to tread lightly on that world, co-creating the wealth we need for the whole of humankind, today and tomorrow.

Hope in Hell is an investigation into what it will take to overcome that kind of cognitive dissonance – to get both feet aligned with the kind of positive vision of which we can now discern more and more different elements. And that alignment process all comes down to hard, clear-sighted and consistent decision-making by governments today.

We do see signs of that. As I'll show in Chapter 4, the reason why the UK leads the world in the deployment of off-shore wind (a technology that is so much more significant on a global basis than most people even begin to understand!) is because the government determined some time ago to support an emerging sector, to take some risks, to use public money strategically, and above all to stick with it.

But that's a rare exception. All too often, it's 'a bit of this and a bit of that', as in phasing out petrol and diesel vehicles by 2030 versus a new coal mine in Cumbria. An 'Active Travel' strategy versus a £27 billion road-building programme. Or it's a small dollop of jam today in the hope of a much bigger dollop tomorrow. The inevitable result is incoherence and no sense of any overarching purpose.

Every single company with a stake in tomorrow's Net Zero Carbon economy has told the Chancellor of the Exchequer, on umpteen occasions, to give them both clarity and consistency. But six months away from that all-important Climate Conference in Glasgow at the end of 2021, we still don't know the direction of travel. We could be on the verge of some kind of 'Roaring Twenties' rebound recovery, ramping up purchasing power at every turn regardless of environmental and climate consequences. Alternatively, we could be on the verge of some kind of 'return to austerity' as the deficit hawks in the Conservative Party ratchet up their anxiety at rising levels of debt.

Or we could still avoid both of these nightmarish outcomes by doubling down on the kind of zero carbon, nature-restoring economy that would both address short-term economic challenges while securing longer-term prosperity across generations.

A just, compassionate, zero-carbon narrative – as high-lighted throughout the following chapters – has to be the best possible way of restoring jobs and injecting purchasing power back into our shattered economies.

After everything we've learned through the COVID-19 crisis about the power of community, solidarity and empathy, there's no reason to suppose that people will unthinkingly re-embrace the false promises of individualistic, me-first consumerism. Or be prepared to put up with austerity-obsessed measures that hit the poorest hardest. Or beg to have their lives blighted all over again by filthy streets and foul air.

But all of this depends on political will, with all of us, in whatever way we can, seizing the moment to avoid yet another 'tragedy of the horizon'. And that means bringing as much pressure to bear on our politicians as we possibly can.

Jonathon Porritt
6 April 2021

Part 1

REASONS TO BE CHEERFUL

1

This Is Personal

'Hope is an embrace of the unknown and the unknowable, an alternative to the certainty of both optimists and pessimists.'

Rebecca Solnit

I wrote the first chapter of *Hope in Hell* in July 2019, on holiday in Cornwall, in the same place we go every year, with a wonderful view of one of Cornwall's oldest and most successful wind farms. The Arctic was on fire. A smoke cloud the size of the whole of the EU was drifting over Siberia. Nearly 5 million hectares of forest were burned. It turned out that July was the warmest month on Earth since we humans first made our mark more than 200,000 years ago.

I wrote the last chapter of *Hope in Hell* in January 2020, with bushfires in Australia still raging, destroying 100,000km² of forest, causing the death of thirty-three people and around 3 billion wild animals, and direct and indirect economic damage more than $100 billion.

Disasters of that kind didn't come as any kind of surprise

(I joined the Green Party back in 1974, and first started campaigning on climate issues with Friends of the Earth in the 1980s), but they now overshadow every other aspect of the work I do today to help build a more sustainable world.

I have absolutely no reason to doubt the warnings of scientists that we have no more than a decade to avoid the horror story of what is referred to as 'runaway climate change' – when natural systems start shifting so fast that there's nothing we can do to stop things getting worse and worse. I see in those warnings a moral imperative that now affects each and every one of us: whatever we can do to avoid that horror story, each in our own way, then we *must* do it.

I was hugely influenced early on in my life by a book called *Blueprint for Survival*, published by *The Ecologist* magazine. In a few short chapters, it explained why a model of progress based on more and more people demanding more and more, every year, on a finite and fragile planet, could only end in tears. As it happens, I didn't actually think that the survival of the human species was seriously at risk when I read it; I just thought that we were managing our affairs, even then, in grotesquely inequitable ways, and that we were trashing the environment in a way that could barely be believed. In short, I understood then that the price we were all being asked to pay (individually and collectively) for 'progress' was bordering on the insane. Fifty years on, I still feel the same. But now I know that the survival of the human species is indeed at risk. That's what the scientists mean when they talk about runaway climate change as 'an existential threat to humankind' – a threat to our long-term survival as a species.

Runaway climate change is undoubtedly a hellish prospect.

Despair often beckons. But I'm also – strangely, and rather wonderfully – resolutely hopeful in a way that I haven't been for a long time. Not only do I believe that *all* the solutions we need to address the Climate Emergency are *already* to hand, but I now see all around me the stirrings of an unprecedented economic and political transformation that will help us avoid that nightmare. What's more, that transformation could, in the not too distant future, lead to the creation of a world that I believe the vast majority of us yearn for – a world in which each of us gets a chance to be brought up lovingly, to be properly educated, to work hard and fulfil our true potential; in which today's astonishing wealth is shared so much more fairly; in which we feel secure in our communities, with decent homes and healthcare, with an expectation of being cared for as we get older; in which our local environment is kept clean, green and healthy, and the global environment is made secure for our children and all future generations.

That's why I'm now so improbably hopeful. Our world – our moral universe – has suddenly become binary. There is no moral universe – in any country, under any faith system, whatever one's personal situation or political beliefs – in which 'business as usual' can possibly provide any rationale for us continuing to live our lives in the way that we do today. COVID-19 provides ample confirmation of that reality. Either we start doing *everything* we can to help make the future as good as it still can be, not just for our children and grandchildren, but for the whole of humankind. Or we don't.

I spent the first twenty years of my life (in the Green Party and Friends of the Earth) campaigning tirelessly to force environmental issues onto the agenda, to persuade politicians and business leaders to rethink their environment-trashing policies and practices. Some campaigns we won; most we lost. Some ideas of ours were surreptitiously taken up by mainstream politicians; most were ignored. For the better part of two decades, it was a war of attrition, and there came a point for me personally (inspired by two weeks at the Earth Summit in Rio de Janeiro in 1992) where the emotional repertoire I was playing with during those years – anger, fear, blame and guilt – was completely exhausted.

So I came back from Rio determined to find a way of working with people's positive energy, to harness the new awareness emerging at that time – particularly within progressive companies. In the next couple of years, together with my Green Party colleagues Sara Parkin and Paul Ekins, I co-founded Forum for the Future (happily, still thriving, and now established in New York, Singapore and Mumbai as well as in London) and helped set up the Prince of Wales's Business and Sustainability Programme (happily, still thriving as part of the hugely influential Cambridge Institute for Sustainability Leadership). Some of my fellow Greenies gave me a very hard time back then for our readiness to work with the private sector, but I felt it was the right thing to do at the time, and I stand by that judgement now. It's still hard to imagine how we're going to transition to a peaceful, just and genuinely sustainable world without progressive companies playing a big part in that process.

Beyond that, as chair of the UK Sustainable Development Commission (SDC), set up by Tony Blair in 2000, I spent nine years of my life working with a dozen Whitehall departments, to ensure 'that sustainable development should become the central organising principle of Government'. Despite all the bureaucratic barriers, we made good progress during that time – very good progress with some departments – only to see both the Commission and all that work unceremoniously swept aside in 2010 by the Conservatives and the Lib Dems. That was a bitter blow – compounded by the fact that Labour, in opposition, did next to nothing to keep alive the idea of sustainable economic development, leaving only the Green Party to go on telling people the real truth of what was happening to the world, to our climate, and to any semblance of a fairer, more compassionate world in the future.

And then came the much-hyped Paris Agreement, signed up to by (almost) all nations back in 2015. I wasn't in Paris for that breakthrough moment in climate diplomacy, but I remember all too well the twenty-four hours between the near-euphoric mixture of joy and relief as the media reported on world leaders' commitment to doing 'everything in their power' to avoid runaway climate change, followed by the dawning realisation just a few hours later that what had been signed up to (in terms of all their country-specific commitments) would actually take us careering past a civilisation-threatening average temperature increase before the end of the century. The kind of hope that sustained me at that time (and had done so, more or less, over the previous forty years) shrivelled in my soul at that moment. Since then, I've known deep down that things would almost inevitably come to this time of reckoning.

All of which means, in a way that is both incredibly simple and incredibly complex, that I have to move on. I have to use whatever influence I may still have to persuade people that *there is no hope whatsoever in another ten years of incremental change.*

If we're to do what we need to do to avoid runaway climate change, we have to force our politicians to step up and do what is now needed. I've come to the conclusion that we have no choice: without mass civil disobedience, at this very late stage, I cannot see any other way of avoiding that threat of runaway climate change. And I cannot, in good faith, advocate for that kind of game-changing civil disobedience without being prepared to be part of it myself, in one way or another.

Our two daughters are thirty-two and twenty-nine. For both my wife and myself, the privilege of being a parent has been and still is beyond words. It has helped make us who we are today, just as I hope our lives have helped make our daughters who they are. And, today, they're good – each in her own way (I write this with their permission!) with a pretty conventional mix of hopes and fears about jobs, personal relationships, the future and so on.

But what of that future? I was roughly their age back in 1980, six years into a career as a teacher in a London comprehensive, and nearly ten years into my somewhat improbable life as a Green activist. All being well for them personally, they'll be my age around 2060. On a 'business-as-usual' basis, the world in 2060 – my daughters' world in 2060, your children's world in 2060, your grandchildren's world in 2060 – will already be the closest approximation to hell on Earth that you can possibly imagine, because of runaway climate change. (I spell out those reasons in Part 2.)

That realisation gives rise to every emotional response you can possibly imagine: regret, grief, guilt, anger, rage – and the rest. But threaded through all of that is what I can only describe, in a rather banal way, as 'just-in-time hope'. Hope born of today's incontrovertible climate science. Hope born of an enduring belief that, on balance, most of us want to live our lives guided by the 'Golden Rule' – 'treat other people as you yourself would hope to be treated by other people'. Hope inspired by all those millions of people, young and old, now actively demanding radical change before it's too late.

For me personally, at this late stage, that means not just pushing 'the solutions agenda' harder and harder, but actively supporting the use of civil disobedience, encouraging more and more people to be prepared to break the laws of their land, peacefully and non-violently, to oblige politicians and decision-makers to change those laws before it's too late. I see the school strikes movement (which goes under many different names around the world) as an inspirational manifestation of civil disobedience on the part of young people, purposefully absenting themselves from their schools in order to remind politicians of their duty.

For me personally, that means doing everything I possibly can to support young people in their own efforts to build the kind of civilised, compassionate, just and sustainable world that we ourselves have so signally and so immorally failed to deliver.

So much has been written about climate change over the past couple of years, but I hope that the basic narrative of *Hope in Hell* makes it a little bit different:

It is *not* too late to avoid runaway climate change. But it soon will be.

This means we must continue to spell out the uncomfortable implications of today's climate science, while seeking solutions in three areas:

- Radical decarbonisation of the economy through technology, stopping CO_2 and other greenhouse gases getting into the atmosphere;
- Radical recarbonisation of the natural world through biology, taking CO_2 back out of the atmosphere;
- Radical political disruption through civil disobedience.

That is the only appropriate response to today's Climate Emergency.

And that is the only place where authentic hope is to be found.

I shall return to the implications of this analysis, for each and every one of us, in the final chapter.

2

THE POWER OF HOPE

'It always seems impossible until it's done.'

NELSON MANDELA

Despite having lived my entire adult life witnessing a progressively worsening ecological crisis, I've always been more drawn to narratives of hope than to the drumbeat of despair. In 1979, I wrote the general election manifesto for the Green Party (then the Ecology Party) as an upbeat declaration of faith in the power of Green politics. When I was director of Friends of the Earth, we made a point of celebrating all the breakthroughs and success stories going on during the 1980s. For the Earth Summit in Rio de Janeiro in 1992, I initiated the Tree of Life project, securing more than a million environmental pledges from people all around the world. Sara Parkin, Paul Ekins and I then set up Forum for the Future in 1996, specifically to inspire both the private and the public sectors with the power of solutions. And I wrote *The World We Made* in 2012 to conjure up a positive vision of what a fair, compassionate and genuinely sustainable world could look like in 2050.

I couldn't possibly have stuck at it over all those years without being able to drink deeply and regularly from a constantly replenished reservoir of hope. I'm drawn to people whose lives are driven by that same impulse, often facing extraordinary challenges in their work. Time after time, I've seen how hope provides some immunity against adversity. All progressive social movements are nurtured by a similar kind of hope, frequently in the face of ridiculous odds. Nelson Mandela's reflection that 'It always seems impossible until it's done' provides reassurance to countless causes, organisations and communities endeavouring to protect what is valuable to them or seeking to make life better for others.

All that said, I would never describe myself as an optimist! There are just too many powerful, destructive forces at work in the world, and too many bystanders reluctant to do anything about those destructive forces, for there to be any justification for optimism. But nor would I describe myself as a pessimist. There are just so many powerful, positive forces at work in the world, with more and more people prepared to stand up for them, for pessimism to make any more sense than optimism. As the acclaimed US author Rebecca Solnit says:

Hope is an embrace of the unknown and the unknowable, an alternative to the certainty of both optimists and pessimists. Optimists think it will all be fine without our involvement; pessimists take the opposite position; both excuse themselves from acting. It's the belief that what we do matters, even though how and when it may matter, who and what it may impact, are not things we can know beforehand.

There's a particular variety of what I call 'shiny optimism' that I find especially irksome – and Ola Rosling (son of the redoubtable Swedish academic Hans Rosling) is one of the shiniest. There's a huge amount of really valuable material in his bestselling book *Factfulness*, but his somewhat selective use of statistics can leave readers with an inaccurate picture of the world as it really is. He correctly points out, for instance, that the share of the Earth's land surface protected as national parks or reserves increased from next to nothing in 1940 to 14.7 per cent in 2016 – but without providing the necessary counter-balance that a huge number of these protected areas are being systematically encroached on by illegal settlers and developers (especially in countries where the population is still increasing), let alone that the rest of the natural world is pretty much collapsing all around us, with more species becoming extinct every year.

In the same way, he celebrates the increase in cereal yields from 1,400kg per hectare in 1961 to 4,000kg in 2014 – without mentioning that this massive 'improvement' in productivity has only been secured by drenching hundreds of millions of hectares in toxic chemicals, completely disrupting the nitrogen cycle through the excessive use of artificial fertilisers, entailing horrendous damage to soil quality and biodiversity, and causing the emission of billions of tonnes of greenhouse gases – year after year after year.

Ola Rosling would no doubt dismiss me as someone in thrall to what he calls 'the negativity instinct', noticing the bad more than the good, misremembering the past, and so on. But 'selective reporting by journalists' and the focus on selective statistics (without appropriate context) is an important part of the reason why vast numbers of people are still going about their daily lives thinking all is well with the world, mindlessly celebrating the wonders of modern technology without understanding any of the downsides.

One of the principal justifications for adopting a more optimistic position than reality might warrant is that this is held to be a more effective way of persuading people to get involved in environmental causes or in campaigns to reduce emissions of greenhouse gases. There's a strong body of work that demonstrates that people who feel optimistic about the future are more likely to take action to help improve it. For instance, telling people that the use of coal, the dirtiest of fossil fuels, is slowly declining provides a much more positive measure than telling them that nearly 40 per cent of the world's electricity is still generated in coal-fired power stations. On the other hand, another body of work shows that people are more likely to be motivated to act by hearing it as it really is, even if that makes them more apprehensive or fearful. It's clearly a fine balancing act that's required.

Putting the optimism/pessimism spectrum to one side, there are many different shades of hopefulness within the climate debate. Many feel very strongly that without hope it's impossible to overcome the sense that many people have that they themselves can't make a difference anyway, so why bother; impossible to inspire people to become active themselves, rather than sympathising from afar with those taking action on their behalf; impossible to confront what is undoubtedly a very disturbing picture, when it's so much easier not to have to think about it at all; impossible to sustain our own personal energy and emotional resources.

I concur with that view – just try advocating for change in a hope-less way rather than a hope-full way! However, I also have some sympathy with those who argue that you can't just use hope instrumentally in that way if there's no rational pretext for remaining hopeful when looking honestly at what the science is telling us. As we'll see in Chapter 9, the voices of those who believe it's already too late to turn things around are increasing all the time – with all sorts of implications for building a powerful movement for change.

One thing I do know: the idea that it might already be too late to prevent a total climate catastrophe impacts people very differently depending on how old they are. And the fact that there are now more and more young people and children having to deal with that sort of possibility represents an almost unimaginably cruel burden to have imposed on them.

GRETA THUNBERG: 'ACT LIKE YOUR HOUSE IS ON FIRE. BECAUSE IT IS.'

That's what lends the extraordinary story of Greta Thunberg an added piquancy. As she's explained it, she started learning about climate change at the age of eight; as she thought through the consequences of today's business-as-usual model of economic development, it was incomprehensible to her that governments were doing next to nothing to avoid what would otherwise become an inevitable climate catastrophe. At eleven, this made her very depressed. She stopped speaking and ate very little. At that time, she was diagnosed with Asperger's syndrome. ('For those of us who are on the spectrum, almost everything is black and white. We aren't very good at lying, and we usually don't enjoy participating in this social game that the rest of you seem so fond of.') With the help of her parents, she somehow worked her way out of that depression, in part by changing their lifestyle to better fit the science that was still preoccupying her. And then, in August 2018, she took it upon herself, without her parents' permission, to go on strike 'for the climate' outside the Swedish Parliament in Stockholm. 'Why should we be studying for a future that soon may be no more, when no one is doing anything whatsoever to protect that future?' The logic is disturbingly compelling.

Fast-forward a few months to March 2019, when more than 1.4 million schoolchildren were out on the streets in over 2,000 cities around the world – an utterly unprecedented explosion of interest for any socially progressive cause. Greta Thunberg is herself the first to say that's not all down to her, but it almost certainly wouldn't have happened like that without her intervention. This isn't really a numbers phenomenon: it's about a different quality of leadership. She has added a completely new dimension to the notion of 'speaking truth to power', acting entirely authentically in service to something beyond herself.

As such, she's a remarkable role model for young people today – and for people of any age, for that matter. I was twenty-four when I joined the Green Party in 1974, naive, isolated and totally dependent on a handful of game-changing books – including Rachel Carson's *Silent Spring*, Fritz Schumacher's *Small Is Beautiful* and *The Ecologist* magazine's *Blueprint for Survival*. For a long time, I had no go-to role models, and only a very small community of kindred spirits. It wasn't until Petra Kelly exploded into my life in 1979, as a co-founder of the Green Party in Germany, a passionate eco-feminist and peace activist (Martin Luther King was one of her most important role models), that Green politics took possession of my spirit as well as of my brain. Forty years on, I can't help but see in Greta Thunberg the same kind of searing honesty and integrity as I saw in Petra Kelly.

It couldn't be more different today for anyone in their teens or early twenties – which is clearly one of the reasons Greta Thunberg attracts such utterly hateful trolling, mostly from sad and embittered middle-aged men, fearful of everything she stands for and of the so-called 'Thunberg effect'. She's not alone in that, of course: every stand-out leader who happens to be a woman, of any age, is today subject to a similar assault and battery on social media.

There's something else at work here. There's no doubt in my mind that the Thunberg effect represents a genuinely disruptive, discontinuous transformation, way beyond the slow incremental pace of change that has bogged down the climate debate for decades. As the author Joanna Macy puts it: 'Sudden shifts can happen in ways that surprise us; structures that appear as fixed and solid as the Berlin Wall can collapse or be dismantled in a very short time. Understanding discontinuous change opens up a genuine sense of possibility.'

And one of the reasons for that is, I believe, particularly significant. It has long been a bugbear of mine that so many environmentalists remain blind to the grotesque inequality that scars our world, and that so many campaigners for social justice remain blind to the collapse of our natural world. Social justice and climate justice are two sides of the same coin: always have been, and always will be.

Greta Thunberg has come in for some criticism from environmentalists for what they describe as 'a lack of political clarity'. For me, that's a bit rich coming from those who've spent their entire life in a miasma of political confusion, incapable of challenging either structural inequality in the global economy or the idiocy of relying on permanent economic growth on a finite planet. Greta Thunberg seems to make these connections almost intuitively.

CLIMATE JUSTICE/SOCIAL JUSTICE

Climate change has never been 'an environmental issue'; it just got put in that box because environmental scientists and campaigners were the first to raise the alarm back in the 1980s – with great foresight, as it happens, as the evidence wasn't as crystal clear then as it is today. Unfortunately, we've paid a heavy price for that particular pigeonholing, as it's allowed politicians to address climate change with as little *real* concern as they've addressed every other environmental issue over the past forty years. In fact, climate change is a civilisational issue, rather than an environmental issue, going right to the heart of today's growth-obsessed economy, challenging our very understanding of what we mean by progress.

Given where we are now, we clearly need a global movement that is nothing like today's environment movement, but much closer in orientation to some of the struggles around social justice and human rights back in the nineteenth and twentieth centuries – a radically transformative movement inspired by those campaigns that made possible what was previously unthinkable. Author and activist Naomi Klein describes climate change as 'the human rights struggle of our time. It's too important to be left to the environmentalists alone.' I shall return to this crucial challenge in Chapter 18, simply because I see this commitment to the radical transformation of today's economy as the only source of authentic hope for addressing today's Climate Emergency.

Which is one of the reasons why so many climate campaigners are increasingly disillusioned at the whole climate diplomacy process, dominated as it is by the annual 'Conference of the Parties' (COP) as the principal way of driving change.

Year after year, these big climate conferences, with their travelling circus of scientists, civil servants and campaigners, continue to address the challenge of accelerating climate change in a weirdly detached, technocratic bubble of their own making. Political analysis, linking the causes and the consequences of climate change to the excesses of contemporary capitalism, as well as to the global elites that maintain their iron grip on the levers of power, is relegated to the margins, to endless side meetings outside the plenary sessions, often to forlorn demonstrations going on outside the bubble. The mismatch between these theatrical events (necessary though they may be) and today's increasingly bitter struggles on the front line of our disintegrating planet borders on the tragic. Wen Stephenson, a wonderful writer and climate activist in the US, brings it right back to *the essence of what we're all doing here*:

I'm a human being. I'm the father of two young children. I'm a husband, a son, a brother, a citizen. And I'm engaged in a struggle for the fate of humanity and of life on Earth. Not a polite debate around the dinner table, or in a classroom, or an editorial meeting – or an Earth Day picnic. I'm talking about *struggle*. A struggle for justice on a global scale. For human dignity and human rights for my fellow human beings, beginning with the poorest and most vulnerable, both far and near. For my own children's future – but not only my children, for all of our children, everywhere. A life-and-death struggle for the survival of all that I love. Because that is what the fight for climate justice is.

Utopian thinking?

Talking of picnics, the original Earth Day on 22 April 1970 mobilised an astonishing 20 million Americans (roughly 10 per cent of the US population at that time) to take part in demonstrations across America – the largest ever civic event on planet Earth. It undoubtedly helped change things, ushering in an era of unprecedented law-making in the USA – including the Clean Air Act, the Endangered Species Act and the Clean Water Act.

Moments like that can make a real difference. People's hope comes in many different shapes and sizes: some quite abstract and back-of-mind, providing at least some reassurance in such a troubled world; others much more hard-edged, something we *do* rather than *have*. There's nothing quiescent about this kind of hope, and it entails real choices about how we live our lives today – in the certain knowledge that the lives of those who follow us will be significantly shaped by the choices that we make. As Alexandria Ocasio-Cortez says: 'Hope is not something that you have. Hope is something that you create with your actions. Hope is something you have to manifest into the world, and once one person has hope, it can be contagious.'

It was in that spirit that I wrote *The World We Made*, in 2012, as a hopeful message looking back from the vantage point of 2050. Precisely because of the choices people started making (from roughly 2020 onwards), it was possible for me to present the world in 2050 as a much better world than today. Although still seriously impacted by climate change, humankind has avoided the horror story of runaway climate change; we've freed ourselves of the false promises of consumerism and growth for growth's sake; we're well into a decadal programme for restoring the natural world; we lead simpler, fairer, more compassionate and responsible lives. And we're much more politically engaged since the emergence in 2019 of a broad-based protest movement, driven primarily by young people, called 'Enough!' (That's one of the predictions that I *did* manage to call correctly!)

It's crucial that we give ourselves time to imagine what that 'better world' might look like, and to picture ourselves living in that world, so that we can become better storytellers as we confront the Climate Emergency. People need that kind of reassurance that such a world *is* still available just so long as we can break out of today's lethal inertia. That's not utopian; it's a simple recognition of the power of human creativity and imagination. This is how Rob Hopkins, the founder of The Transition Network, captures it:

I meet more people every day who have given up, who are sure it's too late, who have no doubt that the future's going to be awful, worse than now, a slow – or rapid – slide into collapse. These stories can become the wall that separates us from other possible future scenarios, and from the capacity to envision and enact a positive future, one in which we've actually tackled our problems with competence and courage. As the French say: 'You can't catch a fly with vinegar.'

As to the *means* by which we get there, that's a very different story. We know only too well what hasn't worked over the past thirty years, and we know that Albert Einstein is credited with defining insanity as 'doing the same thing over and over again, but expecting different results'. For me, our only hope today lies in recognising the true implications of that all too familiar saying, and committing uncompromisingly to addressing the Climate Emergency very differently through the exercise of our collective political will. Including, as I've already mentioned, the inevitability of many different kinds of civil disobedience.

But what grounds are there for supposing we're going to see such a profound transformation in the politics of climate change? Is this just more 'magical thinking', another example of false hope triumphing over immovable reality? As of now, the word 'mass' is the problem; we certainly witnessed plenty of civil disobedience, back in 2019, but the number of those involved on a global basis were still very small. Like many of my colleagues today, I take some comfort from the work of US political scientist Erica Chenoweth, whose investigation of hundreds of campaigns over the past hundred years or so led to the conclusion that it takes only 3.5 per cent of the population of any country actively participating in protests to achieve significant political change. I shall draw on her work in further exploring the likelihood of mass civil disobedience, sustained over an extended period of time, in the UK, the USA and beyond, in subsequent chapters – bearing in mind for now that 3.5 per cent of the UK's population over the age of fifteen amounts to no more than 1.8 million people.

I started this chapter urging readers to look beyond the usual tropes of complacency-inducing climate optimism, as well as the kind of climate pessimism that reduces people to disempowered bystanders. I understand only too well why it is that most politicians and many campaigners still default to the warm embrace of climate optimism, but we know now that such facile escapism is a betrayal of anyone looking for authentic reasons to be hopeful, and a betrayal of young people in particular. What we can be is resolutely hopeful that we'll make a much, much better job of it over the next thirty years than we have over the past thirty years, starting that process with enough humility to listen properly to the passion and the anger of young people.

It's time to remind ourselves exactly what the science of climate change is telling us, before looking at two hugely important areas of potential hope: in the technological revolution already going on all around us, and in today's totally unprecedented openness to change on the part of more and more people in every walk of life.

3

Making Sense of the Science

'The American constitution says I have a right to life, liberty and property. How am I supposed to enjoy life, liberty and property if, one day, the island I live on will be under water?'

Levi Draheim

It was in 1988 that Jim Hansen, the doyen of today's climate science community, first gave evidence as to the seriousness of climate change in front of the US Congress. The world paused momentarily, a few journalists pricked up their ears, most of whom then moved on, only too happy to leave such a big and difficult topic to the scientists, politicians, experts and campaigners. A handful of scientists (including, as we'll see, those employed by the big oil companies) had been working on climate issues long before that, sometimes referring back to the ground-breaking work of Eunice Foote, an amateur scientist in the US whose experiments in the 1850s foreshadowed the discovery of the Earth's greenhouse effect (her paper for the American Association for the Advancement of Science in 1856 had to be read 'on her behalf' by a male colleague!), or, more often, to the insights of an Irish physicist called John Tyndall. In 1859, he was the first to prove the link between incoming solar radiation and the way in which various gases in the atmosphere, including CO_2, absorbed that radiant heat.

In the 1890s, the eminent Swedish chemist, Svante Arrhenius, was the first person to examine what the consequences might be if the amount of CO_2 in the atmosphere was to double. None of these pioneers influenced 'the mainstream', and it wasn't until 1979 that the first big conference on climate change took place in Geneva. The pace only started hotting up when the Intergovernmental Panel on Climate Change (IPCC) was set up in 1988, taking on all the preparatory work for the UN's Framework Convention on Climate Change, which came into being at the Earth Summit in Rio de Janeiro in 1992. It was ratified a couple of years later. Unfortunately, despite endless conferences, protocols and agreements after that, nothing much happened, in practice, until 2015. That was the year the politicians (including Barack Obama and President Xi Jinping) finally got their act together to help deliver the Paris Agreement – the single most important moment so far in this sad and sorry story of climate neglect.

THE BASICS

There is now very little controversy about the basic science of man-made climate change. Although the concentration of CO_2 in the atmosphere is tiny (at 0.04 per cent), this still has a powerful effect on the Earth's temperature. The Earth radiates much of the sun's heat back out to space, but there are certain gases up there, including CO_2, that trap a portion of this heat inside the atmosphere. As we started emitting more and more CO_2 (primarily from the burning of fossil fuels and clearing forests), that concentration of CO_2 began to rise, absorbing more of the sun's radiant heat, with average temperatures creeping up proportionately. Although scientists love to argue the toss on this, the consensus is that the average global temperature has increased by somewhere between 0.9°C and 1.1°C since the start of the Industrial Revolution. Most just settle for a round 1°C.

Climate deniers (of whom there are still a handful in the scientific community) love that sort of uncertainty, and use it all the time to argue that the underlying science must be wrong. But it isn't; it's just very complex and unavoidably messy. In fact, the consensus within the scientific community on the basic global warming hypothesis has been rock solid for a very long time. Back in 2004, Naomi Oreskes analysed all peer-reviewed papers on global climate change written between 1993 and 2003, finding that not one paper rejected the consensus that climate change is a man-made phenomenon.

Makes you wonder, really, why we've had such a tortured debate about that rock-solid science for the past twenty years or more. And why the BBC (as the UK's leading 'public service' broadcaster) interpreted its impartiality mandate, year after year, by insisting that every contribution from a bona fide climate scientist had to be matched by contributions from a cohort of noisy ideologues with barely a smidgeon of scientific expertise between them. Eventually, the BBC came to see the error of its ways just four years ago, but by then the damage had been done. Until recently, there were still significant numbers of people in the UK believing there are legitimate doubts about the scientific consensus on climate change.

Things have been even worse in the USA, where the mainstream media have totally failed in their responsibility to present the science of climate change (see page 63), with the fossil fuel companies spending billions of dollars to secure their influence with a whole generation of politicians system-atically corrupted by those companies' contributions to their election campaigns.

In October 2018, the IPCC brought out a Special Report ('Global Warming of 1.5°C') that shocked the world: for the first time in its unavoidably compromised existence (as a UN body, IPCC reports are almost always out of date, not reflecting the latest science, and always 'fudged' in one way or another to placate the sceptics), it spelled out what's *really* going on. Acting on a mandate from the Paris Conference in 2015, it set out to demonstrate how much worse it would be for the world if the average temperature were to increase by 2°C by 2100 rather than 1.5°C. That mandate came about at the very last moment at the Paris Conference when a number of small island states pointed out that anything close to a 2°C rise would mean their nations would be completely submerged. Hence the rather clumsy wording in the Paris Agreement: 'To hold the increase in the global average temperature to well below 2°C above pre-industrial levels, and to pursue efforts to limit the temperature increase to 1.5°C above pre-industrial levels.'

It still seems difficult to imagine that just half a degree can make that much of a difference. When one's only experience of temperature change is tweaking the thermostat, or seeking out sunshine abroad, it seems improbable that the future of life on Earth (not just for us but for all life forms) depends on a mere 0.5°C. But it really does, as I'll explain later.

What made the Special Report all the more impactful was the uncomfortable reminder that the Paris Agreement wasn't exactly the great triumph that everyone felt obliged to present it as at the time.

As I mentioned in Chapter 1, it quickly became clear in the cold light of a post-Paris dawn that even if *all* the voluntary National Commitments signed up to in Paris in 2015 were actually delivered on (a highly questionable proposition anyway), that would still result in an average temperature increase of more than 3°C by the end of the century. And that's a truly horrendous prospect: at 2°C, things are going to be utterly hellish; at 3°C, it's essentially game over for human civilisation. Which meant, confusingly, that the Paris Agreement was both an incredible breakthrough *and* a death sentence, not just for the small island nations, but pretty much for the whole of humankind.

The authors of that 1.5°C Special Report couldn't possibly have been clearer in terms of telling the politicians what's really happening and how they should respond. Here's my shortened version of their 'Summary for Policymakers':

- Set aside, once and for all, your reliance on that 2°C threshold, on the grounds that it will absolutely *not* ensure any kind of 'safe operating space' for humankind in this, let alone the next, century;
- Understand that the difference between staying below 1.5°C rather than 2°C is not a small matter. In fact,

it's massive, in terms of impacts on agriculture and biodiversity, sea-level rises, loss of reefs and other critical ecosystems, water stress, extreme weather events and so on;

- Accept that the science has moved on, and that the 1.5°C threshold must now be used as the basis for *all* policy responses;
- What that means in practice is that global emissions have to peak by the end of 2020, and reduce by half by 2030;
- And that means richer countries will need to go further and faster than that to ensure that poorer countries are not further disadvantaged in addressing the Climate Emergency.

WHERE DO THE EMISSIONS COME FROM?

The body of the IPCC's Special Report provides lots of detailed pathways and scenarios to demonstrate different outcomes depending on how fast emission reductions can be made. The principal focus is on reducing emissions from energy and industry: from the power sector (both electricity and heat), from transportation, from the built environment, and from carbon-intensive industrial sectors like steel, aluminium and concrete. Between them, these sectors account for about 75 per cent of total greenhouse gas emissions.

As we'll see in the next chapter, the pathway for the power sector is today's equivalent of a climate no-brainer. At the moment, about 25 per cent of the electricity we use around the world comes from renewables (see page 45). But here's the best sentence you'll be reading all day: we already have *all* the technology and *all* the investment capital we need to double that in the next five years, and get close to 100 per cent by 2035. All that stands between us and that trail-blazing break-through are the politicians, many of whom are still marooned in the Age of Fossil Fuels.

With heat (for use both in buildings and industry), it gets a little harder. Decarbonising technologies exist, but they tend still to be more expensive; building and appliance standards are nothing like as demanding as they need to be; and energy efficiency is still seen by most politicians as a 'nice to have' rather than the absolute cornerstone of everything we have to do.

Transport accounts for around 15 per cent of global greenhouse gas emissions, with fossil fuels still providing almost all the requirements for ground-based transport, shipping and aviation. It's here that we're going to have to see some of the biggest changes, driven through by a combination of new technology, market mechanisms and a radically different approach to urban planning.

In all these areas, we need to go after the quick wins and the low-hanging fruit that is still out there, while acknowledging that new policy (based on a combination of regulation, incentivisation, pricing and even 'nudging' to bring about individual behaviour change) can take many years to achieve substantive outcomes. And that's particularly the case when it comes to thinking about emissions from land use, food and farming.

It's understandable that the focus of policymakers has until now been on the 75 per cent of total greenhouse gas emissions that come from energy and industry. But what's become clearer and clearer over the past few years is that we now need to pay equal attention to the remaining 25 per cent of emissions that come from agriculture, deforestation and wider land use changes. In 2017, a paper from The Nature Conservancy, one of the leading NGOs in the US, on 'Natural Climate Solutions', concluded that between a quarter and a third of the emission reductions required under the Paris Agreement could actually be delivered by 're-greening the planet' – through reduced deforestation, forest restoration, peatland and wetland restoration, enhanced sequestration of carbon in the soil, and reduced emissions from both livestock and rice production. As Mark Tercek, former CEO of The Nature Conservancy, said: 'If we're serious about climate change, then we're going to have to get serious about investing in nature. Overall, better management of nature could avert 11.3 billion tonnes of greenhouse gas emissions by 2030, equivalent to China's current emissions from fossil fuel use.'

Interest in this suite of solutions has picked up markedly since the publication of the IPCC's 1.5°C report in 2018, with an understanding that there's a great deal about this land-based agenda that we can get going on almost immediately. This is such an important part of the total solutions agenda that I will keep returning to it in subsequent chapters.

But the great thing is that these two big areas of policy are completely complementary: *decarbonising* energy, transport, industry and the built environment, and *recarbonising* our soils, forests and natural environment. At long last, more and more politicians are beginning to get excited about that combination, recognising the fact that confronting today's Climate

Emergency, as intelligently and purposefully as we need to, will simultaneously bring huge economic, environmental and social benefits.

RESPONSE OF POLITICIANS

Back in 2015, at the time of the Paris Agreement, it was a somewhat messy political scene, with world leaders all over the place in terms of their understanding of the science of climate change. Right up until the last moment, it was by no means certain that the necessary unanimity would be secured (this being a formal UN process), but they got there in the end. Six years on, that world looks very different – and much more divided.

Dangerously divided. The intervening years saw the election of arch-denialist Donald Trump (whose four years in the White House were massively damaging), the election in Brazil of Jair Bolsonaro (the 'Trump of the Tropics'), the re-election of the incorrigibly pro-coal Scott Morrison in Australia, the wilful ambivalence of Justin Trudeau in Canada (signing up to a Climate Emergency one day, and signing off on a multibillion-dollar oil pipeline the very next day), and the problematic loss of the UK as a consistent climate leader within the EU. All of which makes today's 'climate diplomacy' even more daunting than the politics of Paris 2015.

The sole mandate climate negotiators should bear in mind in the run-up to the big climate conference in Glasgow at the end of 2021 are the telling words of Professor Piers Forster, a co-author of the IPCC's 2018 Special Report: 'We have to do everything, and we have to do it immediately.'

So why might anybody suppose that world leaders in 2021 will be any more likely to do their duty to their citizens today than world leaders have been over the past thirty years? Three things. First, not only is the science of climate change now crystal clear, but more and more people, in countries both rich and poor, are already witnessing for themselves the worsening impacts of global heating. Second, the solutions to so many of today's climate challenges just go on getting cheaper and easier, year on year, promising significant economic upsides. Third, citizens the world over are on the move, demanding that their voice be heard and that appropriate action be taken without further delay. There's been little *real* pressure on politicians over the past thirty years, but the growing influence of young people in today's climate justice movement is already changing that particular balance of power in a way that was unimaginable even as recently as 2015.

BUSTING THE JARGON

There's no getting around the fact that the science of climate change is complex. I hope the following 'primer' will be helpful in navigating some of that complexity.

1. THE 'BASKET' OF GREENHOUSE GASES

There are a number of greenhouse gases, all of which contribute to global warming by trapping heat in the atmosphere. CO_2 is by far the most important, but policymakers are also very worried about emissions of methane and nitrous oxide (primarily from agriculture) and certain gases used, for example, in industry and refrigeration. These gases are sometimes bundled together as 'CO_2e' – with that little 'e' telling us that the Global Warming Potential of the different gases has been converted into an 'equivalent' CO_2 measure.

2. WAYS OF MEASURING GREENHOUSE GAS EMISSIONS

2.1 Greenhouse gas emissions by volume

Every year, scientists assess the total volume of greenhouse gases emitted into the atmosphere. Over the last ten years, emissions have risen at a rate of around 1.5% per annum, stabilising only briefly between 2014 and 2016.

The UN Environment Programme's (UNEP) 2019 Emissions Gap Report announced a record figure of 59.1 billion tonnes of CO_2e (including emissions from deforestation and land use change), up from 55.3 billion tonnes in 2018. Emissions for 2020 will undoubtedly be lower because of the pandemic, with early indications showing a drop of around 6.5%. However, scientists do not believe this means that global emissions have now peaked.

2.2 Greenhouse gas emissions by concentration

From 1958 onwards, it's been possible to measure the amount of CO_2 (and other greenhouse gases) in the atmosphere, and show how much they've increased year on year, measured in parts per million (ppm). Back in 1958, there were 316ppm of CO_2 in the atmosphere; in 2017, it was 406ppm; in May 2020, it was 417ppm.

As pointed out by the World Meteorological Organization: 'The last time the Earth experienced a comparable concentration of CO_2 was between three and five million years ago. Back then, the temperature was 2°C to 3°C warmer, and sea level was 10 to 20 metres higher than now.'

2.3 Greenhouse gas emissions by sector and by country

This simply tells us the amount each sector in the global economy, or each country, is contributing to the overall volume of greenhouse gas emissions.

3. USING THE SCIENCE TO MANAGE GREENHOUSE GAS EMISSIONS

3.1 The critical test is how soon we can expect annual greenhouse gas emissions to peak, and how quickly we can then start to reduce emissions, year on year, by sector, by country and so on.

3.2 Because CO_2 remains in the atmosphere for a long time, concentrations of CO_2 and other greenhouse gases in the atmosphere will of course continue to rise, even as the total volume of emissions reduces every year. The ultimate measure of success will be how quickly we can start *reducing* concentrations so as to get back to something around 350ppm.

3.3 Carbon budgets

The links between greenhouse gas emissions, concentrations of greenhouse gases in the atmosphere, and the impact of those emissions on temperature increases, is now really well understood. This allows the Intergovernmental Panel on Climate Change to work out a so-called 'carbon budget': how many more billion tonnes of CO_2 can we afford to put into the atmosphere and still have a reasonable chance of restricting the average temperature increase by the end of the century to no more than 1.5°C? Its headline conclusion: to have a 66 per cent chance of staying below 1.5°C by 2100, our remaining carbon budget is just 420 billion tonnes. 2018's emissions were 55.3 billion tonnes. On a business-as-usual basis, we will therefore have used up that total permissible budget by around 2028.

4. CARBON FOOTPRINTS

All those global estimates and targets don't mean much until scaled down to the different levels where policymakers can have a real impact. There are now well-established methodologies for calculating the combined carbon footprints of whole countries, regions, cities, sectors, companies, factories, organisations, office buildings, hospitals, schools, shopping centres – and, of course, of each and every individual citizen on planet Earth.

5. Restricting average temperature increases

This is all about keeping the increase in the average global temperature (compared with the average temperature at the time of the Industrial Revolution) as low as possible – with an increase of no more than 1.5°C now considered to be the safest threshold for ensuring a more or less stable climate in the future. The higher the temperature increase (2°C, 2.5°C, 3°C etc.), the greater the impact on the planet and on the global economy.

In the words of UNEP in 2019: 'We are on track for a temperature rise of over 3°C. This would bring mass extinctions and large parts of the planet would be uninhabitable.'

4

THE POWER OF TECHNOLOGY

*'The truth is, individuals change their world over
and over, individuals make the future, and they do it
by imagining things can be different.'*

NEIL GAIMAN

THE SOLAR REVOLUTION

Technology will matter as much as anything in the transformation ahead of us. So, let's take heart, right up front, by
acknowledging that many of the technologies we need to
drive this transformation are *already* out there, with a proven
track record, reinforced by a pipeline of new technology and
innovation that will hugely enhance the pace of change in the
future. The solar revolution is perhaps the most remarkable
story of all, and there are now many 'experts' eating great
big slices of humble pie at having so profoundly underestimated this shift. As recently as 2016, Sir David MacKay,
the UK government's former Chief Scientific Advisor to
the Department of Energy and Climate Change (DECC),

described the suggestion that solar, wind and other renewables could power the UK as 'an appalling delusion'. Just two years after that, official figures showed that what is called 'installed capacity' (which refers to the maximum amount of electricity that can be generated from any system under ideal conditions) for UK renewables stood at 42GW, while fossil fuel capacity had fallen to 40.6GW. ('GW' stands for gigawatt. A gigawatt equals 1,000 megawatts. Big power stations come in at anywhere between 1,000 and 2,000 megawatts. Installed capacity is different from the actual amount of electricity supplied to the grid, which is measured in terms of hours of supply – MWh or GWh.)

And we saw the fruits of those investments in 2020, when around 40 per cent of the UK's electricity came from renewables – almost exactly the same as from fossil fuels (with just 2 per cent from coal) and more than twice that from nuclear. The 'appalling delusion' has become an unstoppable revolution. By the end of 2021, it's more than likely that a full 50 per cent of our electricity could come from renewables, and this trend will continue, inexorably, with the last few coal-fired power stations in the UK closing down by 2024, at the same time as further huge increases in offshore wind come online.

However, there are still many people around the world who continue to question whether we're going to see a dramatic acceleration in wind and solar, or just a slow, steady build-up. After all, if you look at the table on page 45, wind and solar combined contributed just 7 per cent of global electricity at the end of 2018.

But here's why things are going to change so much faster

from now on: this accelerated transition is being driven primarily by economics. Between 2010 and 2020, the global average cost of solar photovoltaics (PV) fell by more than 85 per cent. This means that large-scale solar farms are now 'investable projects' in many countries without any subsidy whatsoever. This dramatic fall in prices also makes it so much easier for individuals to get in on the act with roof-mounted solar panels. By the end of 2020, more than 2.65 million households in Australia (around 21 per cent of homes) had invested in solar panels, providing a total of more than 10GW of electricity, the equivalent of half a dozen coal-fired power stations. There comes a point where it simply doesn't matter what ideologically-driven politicians (such as Scott Morrison, the prime minister of Australia, one of the world's most incorrigible defenders of the coal industry) are saying: people not only 'get it' in theory, but are only too happy to take advantage of being able to do something about it *in practice* when it works for them economically. The potential for solar power in Australia has been estimated at more than 170GW, producing more electricity than Australia's total national demand.

Even in the UK, with rather less sunshine than Australia, reliable projections suggest that there will be a huge increase in solar PV and battery storage over the next few years. In December 2019, one of the largest solar plants in the UK came online: NextEnergy Solar Fund's Staughton plant, on a former airfield in Cambridgeshire, will provide electricity for 15,000 households. But what makes it special is that it received no subsidy whatsoever. By the end of 2021, NextEnergy is hoping to have tripled its output. Having acknowledged that solar will be the cheapest generating technology by 2025 (at half the cost of the most efficient gas-fired power stations), the UK Government is now exploring the possibility of a further massive expansion of solar power as part of achieving its target of a 78 per cent reduction in greenhouse gas emissions (from 1990 levels) by 2035.

This is where we are with solar technology *today*. Let loose your imagination and speculate where we might be in 2030. The wonderful reality is that solar power is only just getting going! New materials (including very exciting developments combining standard silicon with a material called perovskite), more efficient manufacturing, and higher conversion ratios will all make a huge difference. Most commercially available solar cells can't do much more than convert around 20 to 25 per cent of the energy from the sun into electricity, but experimental prototypes are already achieving more than 30 per cent. Commercialising that kind of efficiency gain would represent a further massive breakthrough. And not just in the rich world. Because the price of both solar and batteries keeps coming down, year after year, and will do so throughout this decade, solar is rapidly becoming the go-to technology for many poorer countries.

In that regard, what's happening in India is truly astonishing. Even though things have slowed down somewhat over the past couple of years (primarily because of bad policy-making at the state level), India already had 30GW of solar by the end of 2018, well ahead of target; Prime Minister Modi then announced a new target to install an astonishing 100GW by 2022. India is already the lowest-cost producer of solar energy anywhere in the world, primarily because of relatively cheap labour and land, with the natural advantage of 300 clear and sunny days every year. Much of that new solar is providing rural communities across the country with the benefits of off-grid electricity, as well as solar lighting and solar cookers. There's no reason why the use of dirty, dangerous, costly kerosene in rural India shouldn't become a thing of the past within the course of this decade – with huge health and economic benefits for hundreds of millions of Indian citizens.

WIND POWER

Solar has become my go-to cheer-me-up whenever I have to listen to apologists for the fossil fuel or nuclear industries – talking of 'appalling delusions'! But solar is just the start. Next up is wind power. The story here is just as remarkable as the solar story: as the costs come down year on year, the efficiencies increase year on year. It took a very long time for energy experts, planners and politicians to understand just how big wind power was to become, but at long last the boom times for this critically important industry are now here. European countries led the way in 2019, with Denmark relying on wind to provide 48 per cent of its electricity, Ireland 33 per cent, Portugal and Germany 27 per cent. The US total stands at around 7 per cent (with delicious irony, it's Texas that has become the biggest wind-generating state in the USA, providing around 15 per cent of the state's total electricity requirement), and China at around 5.5 per cent. China remains the country where wind power is growing at the most rapid rate. In 2020, it installed 58GW of new wind installations – more than the whole world combined in 2019 – an astonishing figure given that just ten years ago total global installations stood at 39GW!

There is of course plenty of land in both China and Texas – Texan cattle ranchers have been only too happy to discover such a lucrative second income. By contrast, in densely populated countries such as the UK, the debate about 'aesthetics' has been particularly controversial, with a surfeit of lifelong NIMBYs (not-in-my-back-yarders) successfully persuading a Conservative government in 2016 to prohibit almost all new onshore schemes, thereby depriving consumers of by far the cheapest source of electricity in the UK. Despite that barrier, onshore wind in the UK already provides around 8.7 per cent of our electricity, much of it from Scotland.

Offshore wind is very different. With far less opposition to cope with, development over the past few years has been rapid. In fact, the UK still leads the world in terms of the amount of installed offshore wind capacity, representing an incredible engineering success story. In September 2019, a new round of licensing for offshore wind was announced, ensuring that another 7GW will be coming online over the next few years. Costs have fallen so dramatically that this latest round required little subsidy, confirming that offshore wind is now significantly cheaper than nuclear power. Bigger turbines and more efficient high-voltage cables are the principal reasons for this economic triumph, but the industry has also been very successful in reducing high maintenance costs through reliability improvements.

The government has now set a target to go from our

current 10GW of offshore wind energy to 20GW by 2025 and 40GW by 2030 – enough power for every home in the UK. This will require significant levels of new investment, with around £100 million already committed by the government to upgrade key port facilities on the Humber and Teeside. GE Renewable Energy confirmed in February 2021 that it would be investing in a major new factory manufacturing blades for its huge 14MW turbines, 190 of which will be installed at a new wind farm on the Dogger Bank in the North Sea. (This is not quite the biggest turbine in development: as of February 2021, Vestas unveiled plans for a new 15MW turbine – each one of which will be able to provide enough electricity to power 20,000 households!)

Many other countries are now lining up behind this success story, including the US, with its first big wind farm off the coast of Massachusetts coming online in 2022, generating about 800MW of electricity – the equivalent of a large coal-fired power plant. President Biden has committed to doubling US offshore wind, of which the International Energy Agency (IEA) has said there is enough to provide more than double total US electricity demand by 2030. The IEA has also estimated that China's offshore wind capacity will grow from today's 9GW (still just behind the UK) to 170GW by 2040 if it adopts tougher climate targets – as is widely predicted.

So here's what that all means in terms of the contribution of renewable energy to five large industrial economies (2019), as well as to total global electricity supply in 2018:

Country electricity generation total from renewables	US	China	India	Australia	UK	Global total for each technology (5 countries plus the rest of the world)
Solar PV	2.14%	0.21%	3.17%	5.62%	3.91%	2.07%
Wind	6.96%	5.55%	4.14%	6.71%	19.8%	4.76%
Hydro	6.81%	17.81%	10.9%	6.05%	2.21%	16.18%
Waste)	0.4%	0.21%	0.1%	0%	2.75%	4.19%
Biofuels	1.27%	1.41%	2.78%	1.33%	10.15%	1.94%
Geothermal	0.42%	0%	0%	0%	0%	0.33%
Percentage of renewables contribution to total electricity generation	18%	25.2%	21%	19.7%	38.99%	29.28%

Source: https://www.iea.org/ 2019 data except the 'World' column, 2018

At 39 per cent, the UK is doing pretty well. But it's EU countries that are now leading the field, as Professor David Elliott notes:

Germany will soon get around half of its power from renewables, Portugal is already at over 54%, Denmark near 60%, while Sweden is at 66% and Austria over 70%. By 2030 some of these countries could be getting near 100% of their electricity from renewables and should also be beginning to meet significant shares of their heat and transport needs using renewables. Sweden already gets around 54% of all its energy from renewables, Norway and Iceland are both at around 70%.

That 29 per cent of total global electricity supply is pretty impressive when you think that most countries have struggled to get the right kind of policies in place over the past decade, to provide the right kind of incentives and market conditions, against a backdrop of incumbent interests in both fossil fuels and the nuclear industry dominating the political space with their high-level lobbying and readiness to buy off politicians with all kinds of financial support and corrupt deals. Now it's different. Both wind and solar can stand on their own two feet, without the need for continuing taxpayer support – unlike the fossil fuel and nuclear industries, which are still subsidised to the tune of hundreds of billions of dollars every year (see page 229) by many governments. People are waking up to this absurdly wasteful injustice; younger politicians are far less 'invested' in yesterday's energy economy and less dependent on corrupt fossil fuel dollars.

The IEA report referred to above astonished energy experts around the world, demonstrating that offshore wind could theoretically provide enough clean electricity to meet *total* global electricity demand. It predicted that offshore wind will grow fifteenfold to become a $1 trillion industry over the next twenty years due to plummeting costs and new technology breakthroughs – including floating wind farms that can be installed further out at sea.

Floating wind turbines are very much the new frontier in this booming industry, allowing developers to harness stronger wind speeds in deeper waters – anywhere between 60m and 120m. This is going to be particularly important for countries like the USA and Japan. It's also a technology where the big oil companies are likely to become major players, harnessing their offshore expertise and logistical resources. BP, Shell and Total have all confirmed major new collaborations over the course of the last year, both in Europe and in the USA.

Fascinatingly, the IEA report also suggested that cheap offshore wind energy could be used to produce 'green hydrogen' through the electrolysis of seawater. Both the UK (with a pilot project in the North Sea) and Australia are now investing in this potential development. Low-cost, near-zero carbon hydrogen could rapidly become an attractive alternative to gas for use in both heating and heavy industry, especially for Australia as demand for its coal and Liquefied Natural Gas declines in Japan, China and South Korea. Japan hopes to substitute hydrogen for 20 per cent of its current fossil fuel use by 2030, and Australia is very well placed to provide that using its extensive solar and wind resources (see page 40).

Despite its very conservative positioning in the global energy industry, the IEA is gradually ramping up its projections for the share of electricity that renewable sources will provide. In its 'Sustainable Development Scenario' (which assumes that politicians will eventually get their act together), it projects a 52 per cent share for renewables by 2030 and 67 per cent by 2040. In October 2020, the commercial forecast suggested that a surge in new investment will push renewables to a 40 per cent share of the global electricity market by 2030, overtaking coal as the primary means of producing electricity soon after 2025.

Just pause for a second to reflect on the astonishing implications of such a rapid transformation.

Of equal interest is the growing number of studies

demonstrating the feasibility of 100 per cent renewable energy systems (*total* energy, not just electricity), both globally and for individual countries. The most authoritative study for the UK is the Centre for Alternative Technology's 'Zero Carbon Britain: Rising to the Climate Emergency', which shows how we could get to 'Net Zero emissions' by around 2035, creating hundreds of thousands of new jobs in the process.

We shouldn't underestimate that kind of jobs dividend in terms of persuading people of the benefits that will flow from this dramatic transition. Even today, more than ten times as many people are employed in the broadly defined 'green economy' in the USA (around 9.5 million people) than in the fossil fuel industry (less than 900,000). As we'll see in Chapter 10, this already positive economic reality is going to get a great deal stronger as politicians move to implement a wave of Green New Deals.

I have absolutely no doubt that the uptake of renewables around the world is about to accelerate even faster. It's clear that wind will soon overtake hydropower as the biggest single source, with solar not far behind. Lifetime costs of both solar and wind will continue to fall, rapidly knocking out any competition from coal and gas, let alone from nuclear, which will continue to get more and more expensive every year. Storage technologies are improving all the time, with costs of lithium ion batteries plummeting over the last few years as a consequence of growth in EV markets, down from $1,183/KWh in 2010 to $156/KWh in 2019, according to data from Bloomberg New Energy Finance. And BNEF is predicting $100/KWh by 2024 (the price point at which EVs reach cost parity with petrol and diesel vehicles) and an astonishing

$60/KWh by 2030. These improvements will come through further incremental innovation in battery chemistry and significant economies of scale. Huge new investments are now being made in battery storage; breakthroughs of this kind are critical to ensure the cost-effective integration of variable (or 'intermittent') renewable electricity onto the grid.

In just a decade or so, I see no reason why renewable electricity shouldn't grow from around 40 per cent of the UK's electricity supply today to at least 85 per cent by 2035.

ELECTRIC VEHICLES (EVs)

Pretty much everything else depends on us hitting that sort of target by 2035 – for the simple reason that we have to electrify virtually everything we can in order to get to 'a Net Zero economy', including ground transportation (cars, vans and HGVs are still largely dependent on petrol/diesel today) and heating (still largely gas-dependent). There's every chance that by 2030 the internal combustion engine (ICE) will, to all intents and purposes, find itself on death row as electric vehicles and hydrogen fuel cells (for HGVs) take over.

I don't actually own a car and haven't done for the past forty-five years – cycling is my passion. But even I can't help but be excited at the speed of the transition now available to us as literally all of the barriers that once stood in the way of EVs are crumbling in front of our eyes: technology, cost, reliability, infrastructure, and that psychological killer known as 'range anxiety'. All of those obstacles have almost nothing to do with the vehicles themselves, and almost everything to do with battery technology. Back in 2017, a think-tank called

RethinkX challenged politicians and transport experts to start thinking very differently about EVs.

> We are on the cusp of one of the fastest, deepest, most consequential disruptions of transportation in history. By 2030, within ten years of regulatory approval of autonomous vehicles (AVs), 95 per cent of US passenger miles travelled will be served by on-demand autonomous electric vehicles owned by fleets, not individuals, in a new business model we call 'transport-as-a-service' (TaaS). This TaaS disruption will have enormous implications across the transportation and oil industries, decimating entire portions of their value chains, causing oil demand and prices to plummet, and destroying trillions of dollars in investor value – but also creating trillions of dollars in new business opportunities, consumer surplus and GDP growth.

The 2030 target might now be seen as somewhat ambitious, especially as the trigger for this dramatic transition (the widespread regulatory approval of autonomous vehicles) is still some way off. But the signals of this shift are coming through loud and clear. In January 2020, the South Korean car manufacturer, Kia, announced a five-year $25 billion strategy to shift to EVs, aiming for a quarter of its sales to come from EVs and hybrids by 2025. In April 2019, the mayor of Los Angeles launched an ambitious citywide Green New Deal, including a mandate that 25 per cent of all vehicles sold will have to be EVs by 2025, and 80 per cent by 2035.

The arrival of Joe Biden in the White House has moved things on even further. Car giant General Motors promptly announced that it would end sales of all petrol and diesel vehicles by 2035, an astonishing turnaround for a company that had been fiercely resisting improved fuel efficiency standards right up until November 2020. In the words of GM's CEO, Mary Barra: 'President-Elect Biden recently said: "I believe we can own the 21st century car market again by moving to electric vehicles." We at General Motors couldn't agree more.'

The Bank of America has predicted that EVs will account for 40 per cent of all car sales in the USA by 2030: 'EVs will likely start to erode the last major bastion of oil demand growth in the early 2020s, and cause global oil demand to peak by 2030.' The shock to the global economy of the COVID-19 pandemic all but guarantees that this 'peak oil moment' will come a great deal earlier than that.

Cars using petrol or diesel waste hundreds of times more raw materials than their battery equivalents. A new report from the think tank Transport and Environment points to a 'double standard' used when assessing the relative merits of electric and fossil fuel vehicles:

> 'When it comes to raw materials there is simply no comparison. Over its lifetime, an average fossil-fuel car burns the equivalent of a stack of oil barrels 25 storeys high. If you take into account the recycling of battery materials, only around 30kg of metals would be lost – roughly the size of a football.'

According to Bloomberg New Energy Finance in 2019, the 'tipping point' (where the unsubsidised cost of electric vehicles becomes competitive with the cost of internal combustion vehicles) will be in 2024. By 2025, it predicts that EVs will make up 19 per cent of vehicle sales in China, 14 per cent in Europe, and 11 per cent in the United States. The trend is clear. Ford has recently announced that it will be offering only electric vehicles in Europe from 2030 onwards. Jaguar Land Rover made the same pledge for 2025.

And it won't be long before new fast-charging batteries eliminate that last psychological barrier that so many people experience at the thought of having to wait around for hours while their batteries get recharged. A new collaboration between StoreDot (an Israeli company), Eve Energy in China and BP will start mass production in 2021 of a battery that can be fully charged in 5 minutes, raising the possibility that petrol and diesel forecourts will be able to transition seamlessly to all-electric forecourts – 'batteries are the new oil', as it were.

Which makes some commentators much more bullish about the speed of transition, predicting that EV sales will account for anywhere between 20 and 30 per cent of all global vehicle sales by 2028. As author and energy guru Jeremy Rifkin comments: 'At this juncture, we will likely see the beginning of the collapse of the fossil fuel civilization. It should be noted that 96 million barrels of oil are consumed around the world each day, and transport accounts for approximately 62.5% of all the oil used. The numbers speak for themselves.'

More and more people acknowledge that the death of the ICE is now inevitable; all that matters is how quickly and cost-effectively that transition can be engineered. In today's Climate Emergency, it needs to happen just as fast as possible. In particular, we need to see sales of gas-guzzling SUVs start to decline *immediately*; despite stagnation in global car markets overall, sales of SUVs are still growing. Between 2010 and

2018, SUVs doubled their market share (from 17 to 39 per cent), contributing more to the increase in global greenhouse gas emissions during that time than the increases from aviation and HGVs combined. In 2020, 42 per cent of buyers of new cars chose an SUV, which on average consumes 20 per cent more energy than a conventional, medium-sized car. Higher emissions from those SUVs cancelled out all the benefits of significantly increased sales of EVs (up 28 per cent) in that year!

It's clear to anyone who can set aside their fossil fuel blinkers that the power of these technologies (solar, wind, other renewables, EVs, batteries – and all the associated infrastructure that goes with them) is already enabling a rapid and potentially planet-saving transition. The technologies work. They're now completely cost-effective, and already outcompeting fossil fuels on a 'levelized cost' basis – where all the different elements involved in pricing energy are taken into account.

Moreover, they're also much cleaner, ensuring dramatic reductions in healthcare costs as the horror story of urban air pollution, causing millions of deaths every year, becomes a thing of the past. A new study published in Environmental Research in February 2021 showed that the burning of fossil fuels (in power stations, cars and factories) was responsible for an astonishing 8.7 million deaths in 2018, one out of five of all people who died that year. Without those emissions, the report calculated that average life expectancy for people around the world would increase by more than a year, while economic and health costs would fall by about $3 trillion.

Delhi is the world's most polluted city, recording an estimated 54,000 deaths in 2020 – five times as many people died from pollution than COVID-19 in Delhi during the year.

Tokyo saw 40,000 deaths, compared with 1,180 COVID-19 fatalities. And there are now growing concerns that exposure to air pollution significantly increases the risk of infertility.

The accelerated electrification of transport cannot come too fast. As with all technologies, however, there are inevitable downsides. Battery manufacture is a pretty dirty business, and mining all the metals and rare earths needed both for batteries and renewable energy technologies will come with significant environmental challenges. Demand for lithium, cobalt, nickel and certain rare earths could increase tenfold by 2050. A hard-hitting report from the Business and Human Rights Resource Centre in September 2019 revealed the full extent of human rights violations and destructive environmental practices from mining operations in the Democratic Republic of Congo, Zambia, Chile, China, India and Brazil.

A new report from Amnesty International in February 2021 includes a set of high-level principles to ensure that the risk of such human rights abuses are minimised, and includes some particularly powerful recommendations for battery manufacturers to reduce total raw material usage and ensure maximum recyclability of key metals – a process that would be greatly enhanced if the EU and the USA introduced new 'circular manufacturing' mandates. We have to get this right. We have to ensure that these supply chains are completely transparent, free of human rights abuses, with all mining operations based on best environmental practices. This is going to be a huge challenge; it would be a tragedy if 'the age of renewables' ended up with as grim an environmental and human rights legacy as the Age of Fossil Fuels.

And that's why we should already be thinking of EVs as

no more than a temporary stopgap on the way to integrated urban mobility systems, where nobody will ever again need to own their own car. The RethinkX study forecasts that today's emerging car-sharing services will rapidly evolve into large, citywide shared electric vehicle fleets, relying entirely on autonomous EVs, increasing vehicle utilisation rates ten times over – on average, privately owned cars today are driven (rather than parked) for no more than 5 per cent of their life-time. This will dramatically reduce the volume of overall car sales and overall raw material usage.

These are not crazy dreams: these are market trends, impacting on the lives of more and more people every year. The solar revolution provides the foundations for everything else we need to do. There's no shortage of money out there, and as the risks associated with fossil fuels and all carbon-intensive technologies keep on ramping up, more and more of that money will move out of yesterday's money-spinners into today's equivalents. According to the 2019 Global Trends in Renewable Energy Investment Report (from UNEP), $2.6 trillion was invested between 2010 and the end of 2019, with total installed capacity quadrupling during the decade, con-tributing massively to that remarkable 29 per cent of global electricity generation. However, we haven't seen much of an increase in annual investment for a few years now, with it still coming in at around $275 billion per annum. 2018 was a particularly disappointing year, and the pandemic also had a big impact on 2020 investment. To achieve the kind of breakthroughs I've been talking about in this chapter it would require annual investments of around $1 trillion – so there's still a long way to go.

Energy efficiency

A lot of that new investment will need to go into energy efficiency rather than into new supply. Over the years, I've got used to the fact that politicians and investors *never* get as excited about saving energy as they do about generating it, but that too is going to have to change. As economies continue to grow (albeit at slower rates than in the past), and population continues to increase (by around 80 million people a year – a challenge to which I will return in Chapter 17), we can't go on expanding energy supply indefinitely, even if it is all renewable. Renewable energy technologies also depend on finite natural resources – metals, rare earths etc – which have to be mined, which in turn requires a huge amount of energy. We have to reduce the flow of materials through the economy as well as the amount of energy we use. Which means we have to make sure all those green electrons are being used as efficiently as possible in the buildings we live in, the appliances we rely on, in the production of cement, steel and aluminium, and in all manufacturing and mining activities.

So, it's highly significant just how much emphasis campaigners are now putting on efficiency as the foundation for any transformational Green New Deal, both in the USA and here in the UK. Key to developing such policies is the growing realisation that there are only two major labourintensive sources of local jobs: face-to-face caring in health and social services, and infrastructure improvement and retrofitting housing. Such work has the advantage of being hard to automate and impossible to relocate abroad.

With the highest levels of fuel poverty in the EU, much of

the UK's existing housing stock is in an appalling condition, imposing a huge burden on local health budgets. A nationwide retrofit programme is now a critical priority, much of which will have to be done locally, benefiting every city, town and village through new jobs and investment. In the UK, the Green New Deal Group is pushing this 'jobs in every constituency' message very hard indeed, calling for a '30 by 30' campaign – an initiative to make 30 million UK buildings (28 million homes and 2 million commercial and public sector buildings) energy-efficient by 2030. As ever, the jobs angle is absolutely critical.

Sadly, this is an area characterised by staggering levels of incompetence and indifference in successive governments here in the UK. Household emissions from heating and hot water (roughly 20 per cent of the UK's emissions) must reduce by a massive 95 per cent if the UK is to achieve its Net Zero target by 2050. According to the Committee on Climate Change, that will require up to 20,000 homes and other buildings to be retrofitted every week. For the next 30 years. The Committee has also pointed out that the number of homes being insulated today has dropped by over 90 per cent since 2012, owing mostly to the government's axing of Labour's tried and tested retrofit initiatives.

All that Boris Johnson's government was able to come up with, as part of its 'Build Back Better' strategy in 2020, was the Green Homes Scheme, promising grants to individual homeowners to help make their homes more energy efficient. Fifteen months later, the Scheme was abruptly terminated, having failed to deliver on any of its ambitious targets.

Putting such policy failures to one side for the moment, it's impossible *not* to be hopeful about the potential for change from the renewables and efficiency revolution, eventually touching every aspect of our lives while radically reducing emissions of greenhouse gases, in a timeframe that makes the idea of 'net zero prosperity for all' entirely achievable. That, of course, is not the end of it. If it were that simple, we wouldn't find ourselves contemplating the possible collapse of human civilisation. Nor does it mean that technology will sort it all out for us, without us having to fundamentally rethink the nature of wealth creation, the absurdities of today's psychologically debilitating consumerism, and even the meaning of progress itself. But it does underpin what I call 'the doability' of it all, making it entirely rational for all of us to be planning for success rather than wallowing in premature despair.

That's why I've never understood why environmentalists find themselves so regularly accused of being 'anti-technology'. We're *not* 'anti-technology'. But we're certainly 'anti-fantasy'. And there are any number of fantastical techno-fixes (including nuclear power and certain kinds of geoengineering) on which defenders of the old world remain doggedly fixated.

Bill Gates's new book, *How to Avoid a Climate Disaster*, positively bristles with one techno-fix after another. His 'Breakthrough Energy' initiative has invested in more than 40 cutting-edge companies since 2015, including a number of the technologies (like Direct Air Capture to get CO_2 out of the atmosphere) that I refer to in Chapter 12. But he absolutely doesn't get the scale of the renewables revolution, and in the same vein as Dominic Cummings, Boris Johnson's erstwhile special advisor, finds energy efficiency boring. His wording here is such a giveaway: 'I used to scoff at the notion that using power more efficiently would make a dent in climate change. I haven't abandoned that view entirely, but I did soften it when I realised just how much land it would take to generate lots more electricity from solar and wind.' Meaning that the only thing that helped this fabled engineer to get his head around the importance of energy efficiency was the fact that solar power will take up a small fraction of available land!

His biggest techno-fixing passion is for nuclear power, which he erroneously describes as a 'zero carbon option', providing entirely 'carbon-free electricity' – either through ignorance or deliberate mischaracterisation. Back in 2008, he set up a company called TerraPower to investigate the feasibility of a new 'travelling wave nuclear reactor' using molten salt as a coolant. Thirteen years on, he's had to acknowledge that TerraPower is still 'years away from breaking ground' on a prototype reactor. In other words, it's just another pretty standard nuclear fantasy.

I started this chapter with a quote from Professor Sir David MacKay, a brilliant physicist and academic. Right up until his untimely death in 2016, he continued to see renewables as 'an appalling delusion', while remaining devoted to nuclear power and the idea of capturing carbon from large coal and gas-fired-power stations (carbon capture and storage). He was, in effect, completely trapped in yesterday's technological orthodoxies, and correspondingly sceptical about the technologies of the future. As chief scientific adviser at the Department of Energy at that time, this 'incumbent mindset' proved to be highly problematic.

NUCLEAR FANTASIES

Which is why I have to end this chapter by briefly explaining why nuclear power plays no part whatsoever in my vision of an ultra-low-carbon future – not least as I'm astonished at how many people still believe it could and should. Briefly:

- Nuclear power is absolutely not 'low carbon', let alone 'zero carbon'. Once all the life-cycle costs are included (from mining and refining the uranium through to constructing and then decommissioning the reactors and storing the waste), nuclear power is significantly more carbon-intensive than renewable energy.
- On average, nuclear reactors take around fourteen years to build – not much help when we take that 2030 threshold into account. Utility-scale wind and solar farms are up and running in two to three years; rooftop solar in six months.
- After nearly seventy years of operational experience, the nuclear industry still has no idea what to do about the nuclear waste it produces; the costs associated with storing (let alone reprocessing) these wastes are astronomical.
- The links between nuclear power and nuclear weapons are still close, with continuing concerns about proliferation of nuclear weapons in countries that have nuclear reactors.
- Many nuclear reactors (as well as most of those still 'in the pipeline') are sited on the coast in order to take advantage of seawater for cooling purposes.

Conservative projections are now banking on an average sea-level rise of 1 metre by the end of the century, with many experts talking about 2 metres – as I explain in Chapter 7.

• Most well-maintained nuclear reactors, sited in the right sort of location, designed and constructed in the right way, are actually pretty safe. But past experience tells us that there's always another Chernobyl or Fukushima waiting to happen, with horrendous economic and environmental consequences. And that's before we take into account cyber threats and other security risks.

Even if there were answers to all of those problems, the economics of nuclear power condemn this constantly overhyped technology to total irrelevance. As wind and solar continue to get cheaper every year, the costs of nuclear keep on rising. In November 2020, the latest report from asset management company Lazard on what is called the 'levelized cost of energy' (LCOE) assessed the unsubsidised LCOE of offshore wind (at £63 per MWh), onshore wind (at £30 per MWh) and solar (at £27 per MWh) all coming in at a fraction of the cost of new nuclear (at £121). These estimates are very similar to those of Bloomberg New Energy Finance and other independent analysts.

Somewhat bafflingly, however, there would appear to be a growing number of environmentalists ready to cut the nuclear industry some slack. The principal reason for this is, of course, their wholly justified fears about accelerating climate change and about the yawning gap between what the scientists are telling us about the Climate Emergency and the still largely inadequate political response to that Emergency. As we'll see, advocates for nuclear power are vociferously intent on demonstrating that it's now an essential tool in the toolkit for getting that gap narrowed – and for getting us to that distant prospect of 'a net zero carbon economy'. Nothing could be further from the truth.

What's more, impressionistic greenies should be even more cautious regarding the hype about next-generation nuclear – small modular reactors, advanced nuclear reactors, even fusion – back again as the holy grail that has obsessed nuclear enthusiasts since the 1950s!

All such dreams are subject to the same ironclad economic logic. All nuclear (old and new) is impossible without massive government subsidies and provides no net gain for the ultra-low-carbon world we need, while continuing to pose significant and wholly unnecessary risks. Thank heavens we simply don't need it.

So, the mood music on today's low-carbon energy transition has changed – irreversibly. The questions have moved on from 'why?' to 'how?'. Far from there being a residual reservoir of reluctance or indifference, more and more people are now ready for a decade of transformation.

5

THE GREAT AWAKENING

'Another world is not only possible, she is on her
way. On a quiet day, I can hear her breathing.'
ARUNDHATI ROY

At long last, things really are moving. The climate logjam
(with public opinion, politics, science, activists and the media
all more or less stuck in the same place for what seemed like
an age) has well and truly broken.

This has been particularly noticeable in terms of mainstream
media coverage over the past year or so. And that's critical.
According to Media Matters for America, *total* coverage of
climate change issues in 2016 on the four main TV channels
in the USA amounted to precisely fifty minutes across the
entire year: ABC – six minutes; Fox – seven minutes; NBC –
ten minutes; and CBS a massive twenty-seven minutes! That
figure rose to just 142 minutes in 2018 – though none of
the networks' news reports on the fifteen named hurricanes
that hit the USA that year even mentioned climate change.
2019 saw an increase to 238 minutes (up 68%!), but that still

represented only 0.7% of overall broadcast time. However, it's clear that there was a significant uplift in 2020 coverage, particularly through the period of the presidential election.

There are of course many other channels and platforms that US citizens use outside of the major networks, particularly young people. But it's still heartening to see just how high levels of concern and awareness are in the USA despite this historical media failure. In a CBS media poll in September 2019, 28 per cent of citizens questioned considered climate change to be 'a crisis', with another 36 per cent seeing it as a 'serious problem'. Moreover, 71 per cent believe that human activity contributes either 'a lot' (44 per cent) or 'some' (27 per cent) to long-term changes in the Earth's climate. The day of the denialists is long gone, even though up to a quarter of US citizens still think climate change could be a hoax. One particularly important sign of the times is that concern among young people is becoming far less partisan. Another poll in September last year, conducted by Ipsos MORI, showed that 77 per cent of younger Republicans saw climate change as a serious threat – 1 per cent more than Democrats in the same age range. Other polls have shown a huge jump in concern among young Republicans over the past three years.

It's hard to pin this down, but the growing involvement of young people has undoubtedly contributed to the breaking up of the logjam. It's reckoned that more than 7 million schoolchildren and young people protested at least once during 2019, in more than 125 countries. The September 2019 protests were joined by many sympathetic adults and parents. This intergenerational solidarity has become a hallmark both of the school strikes and of other direct action protests, symbolised in the USA by the 81-year-old Jane Fonda being arrested on 5 separate occasions in 2019 in the name of Fire Drill Fridays: 'I will be on the Capitol every Friday for the next four months, rain or shine, inspired and emboldened by the incredible movement our young people have created. I can no longer stand by and let our elected officials empower the industries that are destroying our planet for profit.' And she's been as good as her word since then.

Even more remarkably, the school strikes in September stirred the international trade union movement into significant action. Sharan Burrow, general secretary of the International Trade Union Confederation, called on its 200 million members around the world to stand shoulder to shoulder with young strikers. 'You need to know that your stand is our stand. This is the challenge of our future, and we've been saying for a very long time that there are no jobs on a dead planet.' This was something of a breakthrough moment.

By inverting the 'permission' story, with young people giving adults permission to stand up and do the right thing rather than the other way round, a powerful and (I would argue) unstoppable new element has been added to the climate mix. It's highly significant that both the school strikes and organisations like Extinction Rebellion in the UK, the Sunrise Movement in the USA, and Ende Gelände in Germany, put such strong, consistent emphasis on the importance of science. This has proved to be an effective way of getting beyond the highly politicised arguments around climate change, urging people to make up their own minds as dispassionately as possible about what the latest scientific evidence tells us. Dependence on science allows the clearest possible appeal to people's moral values rather than to any particular political allegiance.

The rapid mobilisation of so many young people's movements – symbolised for me by the image of one solitary young woman, Greta Thunberg, holding up her placard, 'School Strike for the Climate', outside the Swedish Parliament in August 2018, morphing into remarkable scenes of millions of young people on the streets of their cities just year later – is one of the most significant political game-changers I can ever remember. In her speech to the UN Climate Action Summit in 2019, she highlighted the increasingly critical issue of intergenerational responsibility:

This is all wrong. I shouldn't be up here. I should be back in school on the other side of the ocean. Yet you all come to us young people for hope? You have stolen my dreams and my childhood with your empty words [. . .] For more than thirty years, the science has been crystal clear. How dare you continue to look away, and come here saying that you are doing enough, when the politics and solutions needed are still nowhere in sight. [. . .] How dare you pretend that this can be solved with business–as–usual and some technical solutions. [. . .] You are failing us. But young people are starting to understand your betrayal. The eyes of all future generations are upon you. And if you choose to fail us, I say we will never forgive you.

Public attitudes

'Young people are starting to understand your betrayal.' I see this as the first salvo in what I believe will become a long-drawn-out cry of intergenerational rage. It's also one of the principal reasons there suddenly seems to be so much more room for radical new ideas, focused not just on climate change but on a much broader economic transformation – as can be seen in the gathering momentum behind the idea of a Green New Deal in the UK, and even more so in the USA.

The origins of this lie in another young people's organisation, the Sunrise Movement. In the 2018 midterm elections, it set about campaigning against candidates who refused to refuse funding from the fossil fuel industry, and supporting the campaigns of candidates backing renewable energy. One of the candidates they endorsed in the 2018 elections was the formidable Alexandria Ocasio-Cortez (often referred to as 'AOC'), the youngest woman ever to be elected to Congress, who has been working with them since then to promote a Green New Deal for America – focused on 'decarbonisation, jobs and justice'.

(I shall return to various Green New Deal themes in later chapters, but as a huge enthusiast for positive visions of the future, I can't recommend too strongly a quick excursion you might want to take to check out AOC's video, *A Message From the Future*.)

In November 2018, 200 of the Sunrise Movement's activists occupied the offices of Nancy Pelosi, Speaker for the Democrats in the House of Representatives, and a true old-guard Democrat who simply hasn't kept up with how the

world is changing, let alone with the imperative of addressing climate change in far more urgent and radical ways.
Something of that 'intergenerational cry of rage' could be
heard loud and clear in this painful stand-off, summarised in
one placard: 'Dear Democrats: Green New Deal now: step up
or step aside'. For young people like this, the idea that there
are a growing number of adults out there declaring it to be
'too late' to do anything about accelerating climate change (as
we'll see in Chapter 9) is as much a betrayal of their generation
as the betrayal by mainstream politicians so eloquently nailed
by Greta Thunberg in her speech.

So, can young people take any comfort from the latest data
about public concern? Data gathered by Eurobarometer shows
that nearly 80 per cent of EU citizens surveyed see climate
change 'as a very serious problem'; 60 per cent of respondents
said they had personally taken action to fight climate change
over the previous six months – up 11 per cent since 2017.
Basically, both awareness and readiness to take action are on
the up across the EU, and we must hope that will remain
the case here in the UK, even though we are no longer a
member state.

But what of countries like India and China? I endlessly
come across people who believe that concern about climate
change is just 'a rich-world phenomenon', but they're wrong.
In fact, at 71 per cent, India has the highest percentage of
people in any country persuaded that the climate is changing
and humans are mostly responsible, with strong support for
both individuals and government bodies doing a lot more
about it. The percentage in China with the same response to
that question was just 45 per cent, but internet browsing data

from China shows heightened interest in global warming over the past year or so, with 58 per cent of Chinese citizens looking for more action from government.

There's no doubt that awareness *is* gradually converting into greater political will. The Energy and Climate Intelligence Unit maintains a tracker for nations, regions and cities that have declared a Net Zero target. Its update in July 2020 confirmed that about 53 per cent of the world's GDP ($46 trillion) is now covered by existing or planned Net Zero targets. That's a significant advance. But as Greenpeace commented after the European Parliament's climate emergency declaration: 'Our house is on fire. The European Parliament has seen the blaze, but it's not enough to stand by and watch.' It all comes down not just to the percentage of activists involved, but to the number of people who still seem perfectly content just to carry on as 'bystanders', however sympathetic their standing by may be. The 'bystander effect' occurs when the presence of others in an emergency situation discourages an individual from intervening – carrying on as normal because everybody else is carrying on as normal.

BUSINESS MOVES

For the past decade (and, in some cases, for much longer than that) the looming crisis of climate change has commanded more and more attention in board rooms around the world. The year 2015 was something of a turning point in that regard, with leading multinationals playing a highly influential role both in the run-up to and during the Paris Conference. In 2016, the Business and Sustainable Development Commission came together for two years to produce a whole raft of papers staking out the territory for progressive companies around the world. Its flagship paper, 'Better Business, Better World' (2017), provides a compelling case for business engagement, pointing to a possible uplift of \$12 trillion a year and 380 million potential new jobs by 2030.

Many companies are not afraid to use their 'voice' to support more sustainable futures – even in the USA. When President Trump pulled out of the Paris Agreement in June 2017, the decision was met with a barrage of criticism from US companies, including Walmart, Goldman Sachs, General Mills, Walt Disney, Kellogg and Nike. Even ExxonMobil and Chevron said that they thought it was a bad decision; it was only US coal companies that were cock-a-hoop about this particular bit of Trump recklessness. Some of the most outspoken critics were the big IT giants – Apple, Twitter, Google, Facebook and others. And that's not because these tech companies have modest carbon footprints. In fact, when their entire value chains are taken into account (from mining the raw materials, to the manufacture and disposal of devices at the end of their – often ridiculously short – life cycle, plus today's massive ICT networks and server farms, which account for around 45 per cent of the entire ICT footprint), the sector's total footprint is around 4 per cent of global greenhouse gas emissions.

So, whatever we may think about these companies (and I'm one of those who believes that they will need to be broken up and regulated to within an inch of their corporate lives before they take control of even more of our own lives than they already have), their decarbonisation commitments are impressive. Microsoft and SAP have been using 100 per cent renewable electricity since 2014; Google's data centres were powered by 100 per cent renewable electricity back in 2017, and Apple's data centres from 2018. Facebook, Twitter, AT&T, Cisco and many others are making good progress.

Business-led initiatives often take a while to land properly. Launched in 2014 by the World Resources Institute and the UN Global Compact, the Science-Based Targets initiative struggled to get much traction at the start. But at the time of writing there are more than 1,400 companies following strict guidelines to assess their carbon footprint and plan actions to ensure they're in line with what the science is now telling us. The Climate Group's 'RE100' was also launched in 2014, bringing together a growing number of companies committed to securing 100 per cent renewable electricity. At the time of writing, there are now 300 global companies signed up. It all adds up, and confirms that boards of directors of more and more companies feel completely confident that rapid decarbonisation is in the best interests of their shareholders.

2020 turned out to be a particularly dynamic year in terms of corporate climate commitments. The 'We Mean Business' Coalition joined forces with the Secretariat of the UN Framework Convention on Climate Change to launch its 'Race to Zero', which is now supported by more than 2300 companies with 'Net Zero' commitments of one kind or another, alongside 75 of the world's biggest investors and hundreds of cities, universities and other partners. The Mission Possible partnership is focused on the 'hard-to-abate' sectors, including shipping, aviation and steel.

We shouldn't be naïve about all this new corporate activity: at one level, they're just moving with the times, and quite a lot of signatories to these collaborations are 'coming along for the ride', as it were, without any serious intent to change their ways. But it's all part of a diverse and dynamic paradigm shift.

And there's a new leadership cohort in town. In 2019, IKEA (with revenues of $43 billion and 200,000 employees) announced its 'climate positive commitment': 'By 2030, our ambition is to reduce more greenhouse gas emissions than are emitted by the entire IKEA value chain – while continuing to grow the IKEA business. This will require a transformational change for our supply chain and in the way all our products are designed.' Since 2009, IKEA has invested around €2.5 billion in green energy projects.

In January 2020, Microsoft went a step further in announcing a $1 billion carbon innovation fund, promising to remove as much CO_2 from the atmosphere as it has emitted since its foundation in 1975.

But we all know that companies like this and other sustainability leaders like Unilever and Patagonia are really not the problem. In October 2019, analysis from the prestigious Climate Accountability Institute in the USA showed that the top twenty companies involved in extracting and selling fossil fuels since 1965 (when some of them already knew what the consequences of year-on-year increases in the emission of greenhouse gases would be) are responsible for 35 per cent of all energy-related emissions during that time, amounting to nearly 500 billion tonnes of CO_2e. Eight of the twenty are investor-owned; twelve are state-owned. No amount of good corporate behaviour in other sectors is going to compensate for their continuing impact. For the eight investor-owned companies (Chevron, ExxonMobil, BP, Shell, Peabody Energy, ConocoPhillips, Total and BHP, in that order by turnover), that will only happen when investors finally recognise the extent of climate risk in their portfolios.

There are now very clear signals that this is beginning to influence the European-based oil and gas majors – BP, Shell and Total. Under its new chief executive, Bernard Looney, BP has committed to produce 40% less oil and gas by 2030 than it did in 2019, and to sell off $25 billion of oil and gas assets by 2025. It has also sold off most of its petrochemicals division to Ineos for a giveaway price of $5 billion. It plans to increase its capital spend on renewables and other alternatives to around 25 per cent of total capex, which will grow its portfolio from around 3GW today – mostly from its solar Joint Venture with BrightSource – to 50GW by 2030.

Shell's commitments are somewhat less ambitious, with plans to reduce output of oil and gas by about 15 per cent by 2030, and to invest around $2 billion of capex on renewables – versus around $8 billion per annum on new oil and gas. It remains extremely enthusiastic about Liquefied Natural Gas, and about Carbon Capture and Storage (see page 182), but at least – and at last – it's getting the hang of what it might mean to become a genuinely integrated energy company rather than a company still totally dependent on its oil and gas assets. That cannot be said of the US-based majors (ExxonMobil, Chevron and ConocoPhillips), whose gestures amount to little more than glossy green lipstick on some very oily pigs.

INVESTORS

> We have reached the inflection point where, in some cases, it is more cost-effective to build and operate new alternative energy projects than it is to maintain existing conventional plants.

Just let those words sink in for a moment. They first appeared in Lazard's hugely influential report in November 2018 on the 'levelized cost of energy', and have been quoted many, many times since then, revealing as they do the speed with which some established voices in the financial services sector are anticipating the end of the era of fossil fuels.

Sadly, such an opinion is not yet commonplace. The Prudential Regulation Authority at the Bank of England has regularly complained of the lack of readiness on the part of UK banks to understand the full extent of the risks they now face because of accelerating climate change. Mark Carney, the former governor of the Bank of England and now UN Special Envoy for Climate Action and Finance, has been tireless in his efforts to persuade financial institutions to get their heads around today's pervasive climate risk, warning investors as far back as 2014 that 'The vast majority of fossil fuel reserves are "unburnable" if global temperature increases are to be limited to below 2°C.'

Through the Financial Stability Board, he and Michael Bloomberg established the Task Force on Climate-related Financial Disclosures (TCFD) in 2015, providing a voluntary set of guidelines to help investors understand the risk of stranded assets in their portfolios – i.e. fossil fuel assets at risk of being closed down long before the end of their operating life as decarbonisation policies and fiscal measures really kick in. As of April 2021, support for the TCFD had grown to more than 1600 organisations, including some of the world's largest banks, asset managers and pension funds, representing a market capitalisation of over $30 trillion. However, against that welcome shift in awareness must be set the uncomfortable market reality revealed in a devastating report from Rainforest Action Network in 2019 that thirty-three major banks had invested a staggering $1.9 trillion in fossil fuels since 2016. They clearly aren't listening carefully enough to Mark Carney: 'Companies that don't adapt will go bankrupt – without question. Just like in any other major structural change, those banks overexposed to fossil fuels will suffer accordingly.'

But nor are most governments listening carefully enough. As I mentioned in the Introduction, COVID-19 recovery programmes in 2020 were still heavily weighted towards fossil fuels. According to Energy Policy Tracker, G20 countries extended support to fossil fuel companies amounting to more than $170 billion during the year – including $14 billion to coal companies! The US was by far the worst offender, but the UK came in second. In September 2020, it went so far as to sign off on 113 licences for new offshore oil and gas operations in the North Sea, and confirmed in March 2021 that it will allow companies to go on exploring for new reserves just as long as they pass some vague 'climate compatibility test'.

We should take some comfort in the fact that this is an extremely fast-moving area, with increasing numbers of investors choosing to get out of fossil fuels. In September 2019, campaigners at 350.org released a report showing that pledges had been made by asset managers and investors to divest more than $11 trillion from fossil fuel companies. That figure represents around 16 per cent of the total value of global equity markets. The latest update on the GoFossilFree.org website records more than 1,200 divestment commitments, amounting to around $14.5 trillion.

Trends of this kind are closely correlated with an extraordinary demographic shift going on. In January 2020, a survey of more than a thousand millennials across the world, carried out by the deVere Group (with over 80,000 clients globally), showed that 77 per cent of them cite environmental, social and governance (ESG) factors as their *top* priority when considering investment opportunities. As deVere's CEO, Nigel Green, commented: 'Research has shown that investments that score well in terms of ESG credentials often outperform the market, and have lower volatility over the long run. And, importantly, because the biggest-ever generational transfer of wealth – likely to be around $30 trillion – from baby boomers to millennials will take place in the next couple of years, ESG investing is set to grow exponentially in the 2020s.'

All of which raises the question of tipping points in capital markets. There will surely come a point where asset managers (including those managing other people's pensions – an enormous pool of investment capital standing at around $41 trillion in 2018) will conclude that the likelihood of much more decisive government action, plus further shifts in relative competitiveness (as captured in the quote from Lazard at the start of this section), makes the fossil fuel investments in their portfolios just too risky – so why not get out of them before those assets start declining in value and then become completely stranded? I shall return to this theme (and to the crucial role of the insurance sector) in Chapter 15.

The other half of the divest/invest movement is the invest bit. The simple reality here is that we will need many trillions of dollars to be invested in low-carbon energy, transport, infrastructure, manufacturing and land use solutions, with every expectation of safe and stable returns. One manifestation of this is in the growth of 'green bonds', issued either by governments or by blue-chip companies, appealing particularly to investors looking for solid, long-term returns over many years – including pension funds. If one looks at the entire 'sustainable debt market' (bonds, loans and so on), Bloomberg New Energy Finance announced in October 2019 that more than $1 trillion had been issued since 2012. This is growing all the time: in 2012, just $5 billion was issued. In 2020 (in what turned out to be a 'bumper year' for such products despite – or possibly because of – the pandemic), that had risen to $732 billion. And according to the rating agency Moody's, more narrowly defined green, social and sustainability bonds are expected to approach a record of $300 billion in 2021, up from $225 billion in 2020.

TURNING TO THE LAW

So things *are* changing. Both mindsets and money are on the move. The once unthinkable idea that the end of the Age of Fossil Fuels might be *imminent*, rather than a long way off in the distant future, is now being thought through with ever greater clarity and purpose.

An interesting dimension in that shift can be seen in the

extraordinary growth of lawsuits being brought either against governments or against fossil fuel companies, seeking redress for damage already done through climate change or legal mandates to prevent further damage. The Sabin Center for Climate Change Law at Columbia University, New York, has identified more than a thousand climate-centred court cases worldwide, and there have been some significant successes over the past few years. In the Netherlands, a number of organisations representing 866 citizens successfully sued the Dutch government in 2019 for its failure to cut greenhouse gas emissions in line with climate science, and ministers have since had to 'reset' their target. 17,000 Dutch climate activists are now pursuing a case against Shell. In Pakistan, a court ruled in favour of a farmer back in 2015, castigating the government's failure to implement its own climate policy; in 2018, the Supreme Court of Colombia ruled that climate change constituted a direct threat to the rights of young people – and went much further by granting 'environmental personhood' to the Colombian Amazon, declaring that this entire ecosystem had its own rights which needed to be protected. That doesn't necessarily mean those rights *will* be protected, but these are important symbolic victories.

New cases are coming forward all the time – including the case of Sacchi et al versus Argentina et al, brought by Greta Thunberg and fifteen children against Argentina, Brazil, France, Turkey and Germany for violating their rights under the UN Convention on the Rights of the Child. There has been a rapid growth in climate lawsuits in the Philippines, South Korea and Japan, and there are more than 110 cases pending in Australia. As well as around 865 cases in the United States.

Perhaps the best-known case in the USA is the high-profile Juliana v. United States. In 2015, a group of twenty-one young people filed a lawsuit against the US administration claiming that it had violated their rights by permitting activities that significantly harmed their right to life and liberty. In 2016, the District Court of Oregon upheld the claim that access to a clean environment was a fundamental right, and allowed the case to proceed, with a trenchant conclusion from District Judge Ann Aiken: 'Exercising my "reasoned judgment", I have no doubt that the right to a climate system capable of sustaining human life is fundamental to a free and ordered society.'

The US administration (under both Barack Obama and Donald Trump) has done its best to have the lawsuit either dismissed outright or bogged down in endless legal delays. In January 2020, a three-judge panel in the US Ninth Circuit of Appeals ruled by two to one to dismiss the case, despite acknowledging that 'the plaintiffs have made a compelling case that action is needed'. This is not necessarily the end of the road for Juliana v. US, but it is a significant setback.

It is by no means certain this will change with Joe Biden in the White House, but attorneys acting on behalf of the 21 plaintiffs are now planning to take their case to the Supreme Court.

Other cases focus on the work done by the Climate Accountability Institute (see page 72), breaking down that detailed data to ascribe direct financial responsibility to oil and gas companies based on their historical emissions since 1965. The cities of San Francisco and Oakland in California are taking the five biggest oil companies to court, demanding that they should each pay 2 per cent of the costs associated with protecting their communities from rising sea levels in the San Francisco Bay. Journalist Fred Pearce provides another example:

> Estimates by the Pacific Institute, a think-tank in California, suggest that a possible 1.4 metres of sea-level rise along the state's coast by 2100 would threaten property worth $100 billion, two-thirds of it in San Francisco Bay. The institute's report argued that oil companies should be liable for the impact of their product, just as tobacco companies were forced in the 1990s to pay out more than $200 billion in a settlement with smokers.

That comparison with tobacco is beginning to unnerve even the most sanguine defender of fossil fuels. If it can indeed be demonstrated that these companies knew about the consequences of rising emissions back in the 1960s and 70s, then the floodgates of litigation may well be bursting open throughout the decade.

It's hard not to feel buoyed up − properly buoyed up − by what's going on out there, in the courts, on the streets, in board rooms and bankers' brains, in our schools and (some of) our universities, with young and old involved in equal measure. And not before time: the situation on the front line of climate change becomes more urgent by the day.

Part 2

THE CLIMATE EMERGENCY

6

TELL THE TRUTH

*'Anyone who believes exponential growth can go on
forever in a finite world is either a madman or an
economist.'*

KENNETH BOULDING

When Extinction Rebellion took possession of Oxford Circus,
in the heart of London, in April 2019, they used a lovely pink
boat to draw attention to their action. And on the side of the
boat were painted the words 'Tell the Truth'.

At first sight, that doesn't look like too radical a demand. But
in the hotly contested terrain of climate science, 'the truth' has
often been downplayed, obscured or even concealed altogether.
The scientific community itself has often been somewhat reluctant to spell out the implications of what its work reveals. Jim
Hansen has consistently complained of 'scientific reticence',
with too many scientists 'erring on the side of least drama',
while others have gone further in identifying 'semi-censorship'
on the part of some of their colleagues. By contrast, lots of
scientists remain very leery about what they call 'doomism'.

Michael Mann, one of today's most eminent climate scientists, warns against presenting the problems 'as unsolvable, feeding a sense of doom, inevitability and hopelessness'.

Michael Mann has been particularly outspoken about the damage done by today's 'doomists', devoting an entire chapter to this in his new book *The New Climate War*, energetically attacking David Wallace-Wells' *The Uninhabitable Earth* amongst others. 'Climate doomism can be paralysing. It's a brilliant strategy for building a truly bipartisan coalition for inaction.' Extreme doomism certainly gives plenty of ammunition to the likes of Rupert Murdoch's malign, climate-denying media empire.

It's hardly surprising, therefore, that politicians have tended, until recently, to be very cautious in what they're saying, anxious to avoid any whiff of fear-mongering. Even NGOs have seemed to fall in with this 'natural restraint', mindful no doubt of perennially hostile commentators (particularly in the USA, UK and Australia) prowling around just waiting to attack.

Much of that reticence has now disappeared. Once the conservative scientists on the IPCC finally 'let it all out' with its 1.5°C Special Report in 2018, there's been little reason for anyone to hold back on telling the truth – other than a fear that such language may prove psychologically counterproductive and even disempowering. And that means the language that we're using today is at last catching up with the science. Though perfectly acceptable from a scientific point of view, 'global warming' always sounded rather welcoming, particularly for people in cold countries. 'Climate change' itself is studiously neutral – the climate could, theoretically, be changing for the better as well as for the worse. These days, however, we hear more about 'global heating', with both 'climate crisis' and 'climate breakdown' putting a very different emphasis on the kind of change that is now underway. And Extinction Rebellion has dramatically injected the notion of a full-blown *Climate Emergency* into all our deliberations. 'Only from truth can we act.'

Expressions of that truth are now spilling out all around us: gutsy, no-holds-barred books; philosophical treatises; proliferating podcasts; dystopian novels; documentaries of every kind; and more and more brilliant reporting from the multiple front lines of today's climate crisis. It's as if a drought has at last broken, inundating us with all this important material that should have been drip-fed to us over the past twenty years.

But the truth about climate change is not easy to cope with. It's very stark. Very disconcerting. I've been worrying away about climate change since the mid-1980s, and have never shied away from the latest shock-horror scientific report or exposé. But now it's different. Any residual illusions that we still have a reasonable length of time to get things sorted, and that we've made a reasonable start through the Paris Agreement, have been comprehensively stripped away. Much of what needs to be done needs to be done in the next decade. Indeed, we need to halve emissions of greenhouse gases during this decade if we're to stay below that 1.5°C threshold by the end of the century – I keep coming back to that one all-important scientific conclusion to help dispel any false hope that anything less ambitious might still be sufficient. What that means in practical terms is that we need to cut emissions by 7.6 per cent every year throughout the decade.

I've already touched on the psychology of 'telling the truth' about accelerating climate change. There's no doubt that a lot of people do indeed turn away from 'too much doom and gloom', or do feel disempowered, or at least discouraged at the thought of just how bad it's going to get, and just how big a challenge we now face to avert something even worse. It's hard to open a dialogue with people if they're already feeling completely demotivated. But the significant success of David Wallace-Wells's *The Uninhabitable Earth* tells me that things are changing. For the first 140 pages there is literally no respite from a litany of lives and places already devastated by the impacts of climate change today. He is completely merciless in laying out the full tableau of climate impacts, or 'elements of chaos', as he describes them: rising temperatures, flooding, wildfires, extreme storms, water shortages, soil erosion, ocean acidification, dying reefs, air pollution, pests and diseases, loss of forests, nutrient collapse, climate refugees, conflict – and I'll pick up on some of these over the next few chapters.

Part of a greater readiness to 'hear the truth' is that so many more people now have some direct experience of climate impacts for themselves – not just in poorer and more vulnerable countries (for which it's always been assumed rich-world citizens have limited concern), but in our own extended backyards. The idea that people are less likely to engage with the consequences of climate change if they have no experience of them personally is strongly borne out by a particularly revealing survey conducted by Yale's Program on Climate Change Communication. This tested an interesting version of 'the proximity principle'. Scientists are clear that the fire season in the USA is both more prolonged and more intense as a direct consequence of our emissions of greenhouse gases. The survey found that people's readiness to accept that direct link depends on their proximity to these wildfires: 69 per cent of respondents in California and 62 per cent in Colorado went along with it, whereas only 36 per cent of respondents in largely wildfire-free Ohio bought into it. Proximity powerfully influences people's perception of reality.

It's not the purpose of *Hope in Hell* to rehash each and every one of David Wallace-Wells's twelve elements of chaos – that would take up the rest of the book! But in order to get at least a sense of the scale of the challenge we face, I just want to dig down into a couple of areas: the problems associated with extreme heat, and, in the next chapter, the full story about melting ice and rising sea levels. Chapter 8 then gets to grips with the all-important consequences of the interactions between different climate phenomena – those extraordinarily scary 'feedback loops'.

Global heating

In terms of global heating, it's what it says on the tin: things are going to go on getting warmer and warmer until we stop trapping more and more CO_2 in the atmosphere. The sheer scale of that warming is staggering: as Jim Hansen reminds us, if you think of all the extra heat that our emissions of CO_2 have trapped inside the atmosphere over the past few decades, it's the equivalent of 400,000 Hiroshima-sized bombs EVERY DAY.

It's almost impossible to get one's head around this. But

it probably explains better than anything else why literally everything is warming up all around the world. Nineteen of the twenty warmest years have happened since 2000. 2020 and 2016 are the joint warmest years on record. But it's astonishing to think that we'll soon be looking back on these two decades and reminding ourselves how pleasantly cool it was!

Nine out of the ten deadliest heatwaves have happened since 2000, and there's been a fiftyfold increase in dangerous heatwaves since 1980; more than a billion people are already 'at risk' from heat stress, with a third of humankind having to put up with heatwaves for twenty or more days in any one year. The World Health Organization (WHO) estimates that heatwaves already kill an estimated 12,000 people annually across the world, and forecasts that, by 2050, deaths from more extreme heatwaves could reach 255,000 annually.

The records tumble so fast it's almost impossible to keep track of them. The summer of 2018 was the hottest ever, with the mercury reaching an astonishing 53°C in Pakistan, and the Arctic witnessing the highest temperatures it had ever experienced – until 2019, that is, which was even hotter, with temperatures a full 3°C higher than the average. June 2020 saw the highest temperature ever recorded in the Arctic – an astonishing 38°C – and February 2020 witnessed the highest temperature ever recorded in Antarctica at 18.3°C.

But would all this be happening anyway – as part of some long-term, cyclical changes in the climate? One of the most helpful scientific advances over the past few years has been the development of what's called 'attribution science', shedding light on the degree to which any particular extreme weather event may or may not have been 'caused' by accelerating climate change. The World Weather Attribution partnership at Oxford University has suggested that all of these heatwaves, on average, are 'twice as likely' to have happened *because of* changes in the climate caused by the emission of greenhouse gases. And those attributed probabilities will inevitably keep on creeping up.

Cities have a particular problem here, with something called the 'heat-island effect' ensuring higher temperatures than outside of built-up areas. At the moment, there are roughly 400 cities around the world with a maximum summertime temperature of 35°C. By 2050, it's reckoned that number will be around 1,000 cities, with a combined population of more than 2 billion people. It's hard to imagine anything else causing this 'doubling+' phenomenon, other than global heating – and equally hard to imagine the discomfort of having to suffer blistering heatwaves each and every year.

It's more than likely that London will be one of those cities well before 2050. A study published by the Public Library of Science in *PLOS One* in July 2019, looking at 520 cities around the world, suggested that by 2050 London was likely to have the climate of Barcelona, Madrid the climate of Marrakesh, Stockholm the climate of Budapest – and Seattle the climate of San Francisco! And just in case readers in London think that sounds rather pleasant, just remember that Barcelona was so short of drinking water in 2008 that it had to start importing it from France. Catalonia is still a 'water stressed region', with the spectre of chronic drought never far from people's minds.

The economic impact of heat stress cannot be ignored. Overall, by 2050, it's been estimated that losses in productivity are expected to be more than 2 per cent across the world, and as high as 12 per cent in the worst-affected regions of South Asia and Africa – amounting to hundreds of billions of US dollars. Investing in air-conditioning is what most people do, understandably, if they can afford it. It's an extraordinary fact that 328 million Americans consume roughly the same amount of electricity on air-conditioning as the total electricity consumption of 1.1 billion people in Africa. Air-conditioning already accounts for around 10 per cent of global electricity consumption, and projections indicate that there will be four times as many people using air-conditioning by 2050 as there are today. Emissions from the use of air-conditioning could increase by a staggering 90 per cent by 2050. All of which, inevitably, just makes the problem worse in terms of increased emissions of CO_2, increased average temperatures, etc. Round and round we go.

Addressing this challenge is going to be one of the highest priorities in today's Climate Emergency. In January 2019, the so-called Kigali Amendment (part of the Montreal Protocol, to which I return in Chapter 18) came into force, having been ratified by sixty-five countries. This amendment aims to reduce by 80 per cent the use of the principal gases employed today for refrigeration and air-conditioning – called hydrofluorocarbons (HFCs) – over the next thirty years. HFCs took the place of other ozone-depleting refrigeration gases back in the 1980s and 1990s, but unfortunately they just happen to be very powerful greenhouse gases in their own right – thousands of times more potent than CO_2 in terms of their Global Warming Potential (GWP).

Happily, there are other refrigerants with a much lower GWP that can be introduced instead, providing a simultaneous opportunity to redesign air-conditioners to make them hugely more energy-efficient. This one technological 'double whammy' has the potential to avoid up to 0.1°C of warming by 2050, and an astonishing 0.4°C by 2100.

Cities are one thing; rural areas another. Some of the worst increases in average temperatures are going to be in India, Pakistan, Bangladesh and even northern China – places where a high percentage of the population still works outside in agriculture and other rural trades. The concept of 'habitability' kicks in significantly here, with the number of hours that can be worked outside without serious risk to workers' health reducing all the time.

As it happens, however, that is by no means the greatest threat to agriculture from global heating. Many of the most productive areas in the world are already struggling, with topsoil being lost through erosion and poor land management 100 times faster than it's being formed – about 75 billion tonnes of soil are lost every year. Higher temperatures and drought conditions are going to exacerbate that process: if we do end up seeing an average 2°C temperature rise by the end of the century, then most of the Middle East, South Asia and southern Europe, plus large parts of Australia, sub-Saharan Africa and South America, will be suffering extended periods of drought. In some cases, permanent drought.

What all this means for regional (let alone global) food security is impossible to fathom – with more than 800 million people already undernourished today. There's been a huge amount of interest over the past few years in the links between drought, migration and security issues. Between 2006 and 2010, for instance, Syria suffered four consecutive years of drought, with devastating impacts on both herders and small farmers. Three million people were forced into extreme poverty, many of whom migrated into Syria's cities, sparking a lot of unrest and protests – the precursors for the terrible civil war that followed.

This wasn't just a climate-induced disaster. For many years before the drought Syria's rural population had been subjected to disastrous policy shifts, as the government set out to 'intensify its agriculture' in order to produce crops for export, such as wheat and cotton. There was widespread corruption, and systematic mismanagement of water and land resources. So we have to be cautious about directly attributing the civil war in Syria to climate change, not least as neighbouring Lebanon suffered pretty much the same drought conditions without a similar implosion. But the cascade is clear: as temperatures rise, evaporation increases, reinforcing the likelihood of water stress and drought. Prolonged droughts are one of the principal factors behind increased migration in many parts of the world, and, all too often, that enforced migration can exacerbate social tensions and instability.

Scientists now believe that for every degree of additional warming, average agricultural yields are likely to decline by up to 10 per cent. That's bad enough, but the work of Professor Irakli Loladze reveals that higher concentrations of CO_2 in the atmosphere are already having a serious effect on the nutritional quality of most of the world's major crops – grains, soya, corn and rice. All that CO_2 is making the plants bigger (notionally a good thing), but it also makes them produce more carbohydrates (glucose and so on) – which are a problem from the point of view of a balanced diet – and fewer desirable proteins and micronutrients such as calcium, zinc and magnesium, as well as certain vitamins. By 2050, there could be tens of millions of people in developing countries at risk of nutrient collapse – on top of those who do not have enough food or who are already suffering from different kinds of micronutrient and dietary deficiency.

Extreme warming is just one of the dozen or so impact areas that I mentioned on page 86. But I hope that this one alone is sufficient to confirm the utter folly of seeking to continue living our lives on any kind of business-as-usual basis, as well as to reaffirm the need to start dramatically reducing emissions of greenhouse gases. Setting aside the pandemic-induced emission reductions in 2020, we now need to ensure that those emissions stabilise by the end of 2021, and start going down rapidly from 2022 onwards with a view to *halving* by 2030.

To some people, that still sounds 'extreme'. But who are the extremists these days? People who are seeking to tell the truth, however tough a mandate that may be, or those who still seem relaxed about an inexorable march towards climate breakdown? If anyone needs further information to help them answer that question, a quick glimpse into the world of melting ice and rising sea levels should suffice.

7

MELTING ICE AND RISING SEAS

*'The human race is challenged more than ever before
to demonstrate our mastery, not over nature but of
ourselves.'*

RACHEL CARSON

Everything frozen, anywhere on Earth, is melting. It's the combined impact of that melting around the world that is causing climate scientists more concern than anything else, which makes this particular climate disruption the most disturbing of them all.

In this chapter, I'm going to look at the effects of melting ice sheets and the glaciers of which they're made up, and in the next chapter at melting sea ice and permafrost. There's a big difference: every tonne of ice lost from ice sheets and glaciers contributes directly to rising sea levels. Melting sea ice in the Arctic does *not* contribute to sea-level rise (think of it as melting ice cubes in your glass), nor does melting permafrost. But they still have an absolutely massive effect on accelerating climate change – as we'll see.

GREENLAND

In 2019, in one of his more bizarre foreign policy outings, Donald Trump made a serious offer to the Danish government to buy Greenland. Putting to one side the fact that any change in their constitutional status rests with the citizens of Greenland themselves, rather than with the Danish government, what was going on in President Trump's discombobulated brain? Was he perhaps looking forward to a distant future when Greenland will indeed be 'green' again, opening up this huge (and, with just 56,000 people, very sparsely populated) island to all sorts of lucrative economic development? If that was his logic, I wonder what sort of time horizon he had in mind?

For the truth of it is that Greenland's ice sheet *is* melting. And that ice sheet is huge — covering about 1.7 million km², with an average depth of around 2km, amounting to around 2,850,000 cubic kilometres of ice. Or, to put it another way, 2,850 million million tonnes of ice! So whatever happens, by way of melting, it will take a long time. But what climatologists are focused on is the 'tipping point' — the point at which it's too late to stop *all* 2,850 million million tonnes melting away into the ocean over time.

Were that to happen, that's the equivalent to at least a 6-metre sea-level rise.

Until recently, we've been used to thinking of sea-level rise in millimetres, or centimetres at most. According to NASA, today's sea level is just 20 centimetres higher on average than it was in 1900. But the rate of increase has been speeding up: eight of those 20 centimetres have impacted on sea levels since

1992, and some locations have seen as much as a 25-centimetre rise due to natural variations. But now we're all going to have to start thinking about sea-level rise measured in whole metres. And it's more than likely this will be our reality before the end of the century.

The year 2019 was absolutely dreadful for the Greenland ice sheet. The melting started earlier and finished later, with record-shattering temperatures (sometimes more than 20°C above normal) and extreme rates of melting, amounting to around 440 billion tonnes. Half of that came in just one incredible month in July. Although this is not quite the worst year ever (that was 2012), scientists have been astonished at just how bad things are getting in Greenland, and just how fast some of the biggest individual glaciers are retreating. The Helheim Glacier (down in the south-east corner of Greenland) has retreated by a staggering 6 miles since 2005. Another large glacier is retreating at a rate of 20 metres a day, compared with 20 metres a year back in 2013. Impacts of this sort were not projected to be happening until well into the second half of this century.

This accelerated melting is believed to be a key factor behind changes in what is called the Atlantic Meridional Overturning Circulation (AMOC), a huge system of both surface and deep currents in the Atlantic (including the Gulf Stream) that plays a critical part in weather patterns around the world. An article published in *Nature Geoscience* in February 2021 revealed that the AMOC is now weaker than it has ever been over the last Millennium, having slowed down by around 15 per cent. Melting in the Arctic dumps large quantities of cold water to the south of

Greenland, affecting the 'conveyor belt' of the Gulf Stream that brings warm water up from the Equator. Scientists are now concerned that the AMOC could completely collapse, at some point between now and the end of the century, with devastating climatic consequences for both the USA and Northern Europe.

Which takes us back to that tipping point. Some scientists believe that the Greenland ice sheet has *already* tipped, with today's average global temperature increase of around 1°C; others believe it won't happen until we've exceeded 1.2°C; still others believe parts of the ice sheet could still survive even at 1.4°C or 1.5°C. What that tells me is that there is still a lot to play for here. Palaeoclimatologists have revealed that Greenland had a pretty big ice sheet around 130,000 years ago – when average summer temperatures were a lot higher than they are today. As Adam Vaughan of the *New Scientist* puts it: 'Every fraction of a degree of warming avoided could help avert the worst. The climate certainly has tipping points. Can societies have them too?'

ANTARCTICA

If Greenland is big, Antarctica is absolutely massive. I'll spare you all the detailed measures of the amount of ice, and jump to the big reveal: Antarctica has nearly eight times more ice than Greenland, and if every last tonne of ice in Antarctica was to melt away, it would increase average sea level by around 60 metres. Ten times as much as the contribution from Greenland. That would only happen over hundreds of years, but if the warming continues, it *will* happen.

Antarctica is so vast that one really has to think of it as two half-continents: West Antarctica (including the Antarctic Peninsula and the Ross Sea) and East Antarctica, which has most of the continent's really massive glaciers. So far, the Antarctic has not been warming up as fast as the Arctic, which has seen an average temperature increase across the whole polar region of around 2.3°C since the 1970s – at least twice the global average. Though other factors may be involved in this (including cloud cover and weather systems), most of it is down to the albedo effect: as the Arctic's sea ice melts, incoming solar radiation is no longer reflected back off those shiny white surfaces, but absorbed by darker waters or exposed land, further ratcheting up the warming effect. (I'll return to this in the next chapter.) Antarctica has no equivalent albedo effect, sitting as it does in the very cold waters and currents of the Southern Ocean.

Sometimes I really don't know how best to convey the implications of rising sea levels. I've been Chancellor of Keele University for the past nine years, and every year I have to give the 'graduation speech' to hundreds of our graduating students. I feel excited for them, of course, but unnerved on their behalf by what awaits them in a world about to be turned upside down by climate change. In July 2018, I may not have quite got that balance right. Just a few weeks before, a major study looking at melting ice in Antarctica had been published in *Nature*, and it had shocked me to the core. It revealed that the rate of melting in West Antarctica had accelerated three-fold since 2013, which could lead, eventually, to the collapse of the entire Western Antarctic Ice Sheet (including huge glaciers like the Thwaites Glacier), which in turn could result in a sea-level rise of between 3.5 and 5 metres. So I stuck all that in my graduation speech.

Some of the students' parents subsequently suggested that I had no right to dump such an apocalyptic burden on young people celebrating one of the most important days in their lives. I can understand that, but I also wanted to use that report to fire them up! As one of the report's co-authors, Professor Martin Siegert from the Grantham Institute in London, put it: 'Some of the changes Antarctica will face are already irreversible, such as the loss of some ice shelves. But there is still a lot that we can prevent or even reverse. And that largely depends on choices made over the next decade.' So, is that an apocalyptic downer or an out-and-out message of hope? I guess it depends on the way you want to hear these things, but it's fair to say that *Hope in the Antarctic* is a documentary title that you're unlikely to hear anytime soon.

That's partly because of the Ross Ice Shelf, and partly because of what's now emerging on the other side of the continent, in East Antarctica. The Ross Ice Shelf, at the bottom of Antarctica as you look at it on a map, is vast – roughly the size of Wales. In places, it's 100 metres thick, with 90 per cent of that ice below sea level. So it's important not just in its own right, but because it acts as a kind of mega-plug holding back many of West Antarctica's biggest glaciers. An ambitious study in April 2018, looking at the north-west sector of the Ross Ice Shelf over a period of four years, revealed that it is now melting ten times faster than had previously been assumed – largely, it seems, because of inflowing warmer water from outside the ice shelf. Not good news.

Thousands of kilometres away, in East Antarctica, things

are also looking rather more wobbly than they did a few years ago. Until recently, the 'prevailing assumption' of most Antarctic scientists was that the glaciers of East Antarctica were as solid as the glaciers of West Antarctica were slushy – indeed, the record shows that East Antarctica was actually increasing by around 5 billion tonnes of new ice every year between 1992 and 2017. Any worries that people had were focused on the massive Totten Glacier (by 'massive' I mean that 3.4 metres of potential sea-level rise are locked up in this one glacier), which was known to be losing ice for more than a decade. But a new report from NASA in December 2018 showed that a number of other glaciers adjacent to Totten were also losing ice, by up to 3 metres a year. The study's headline was that East Antarctica was now losing up to six times more ice than back in 1979, with the suggestion that this might also be attributed to warmer ocean temperatures – which has been the big problem for a long time on the other side of the continent.

There's one final factor to be taken into account before trying to work out just how scared we should be about all this. These vast oceans into which we're now releasing hundreds of billions of tonnes of melted ice are themselves much more sensitive than we might imagine. A full 90 per cent of the excess heat that has not been reflected back into the atmosphere (trapped by all the greenhouse gases we've emitted) has been absorbed by the oceans since 1970. As the oceans warm, they also expand. This phenomenon of 'thermal expansion' has contributed about half of all sea-level rise observed so far, making it a very significant 'threat multiplier' for the future.

Mountain glaciers

Antarctica has fifty times more land-based ice than all moun-tain glaciers in the world combined. But those mountain glaciers are now *all* at risk. Some glaciologists believe we will see the end of the Earth's non-polar ice soon after 2100, and maybe even earlier, unless we drastically cut greenhouse gas emissions over the next two decades. Since glaciers contain around 70 per cent of all fresh water on the planet, that would be devastating for the hundreds of millions of people who depend on that meltwater – even if the contribution to sea-level rise would be very small in comparison to Greenland and Antarctica.

In some parts of the world, it's all happening a lot faster than a 2100 date might imply: in East Africa, where ten of the eighteen glaciers that provided precious water are already gone; in Alaska, now losing around 75 billion tonnes of ice per annum; in Montana, where all 150 glaciers in the Montana Glacier National Park will be gone by 2030; in Peru, where glaciers have already lost nearly 40 per cent of their total mass; in the Alps, where it's 50 per cent. And so on.

Scientists are particularly focused on the Himalayas, some-times referred to as the 'third pole' because of the huge amount of ice across the entire region. A report published in *Science Advances* in 2019 showed that Himalayan glaciers were melt-ing twice as fast between 2000 and 2016 as between 1975 and 2000, coinciding with a significant increase in average tem-peratures across the region. This is a huge worry. Depending on how you do the calculations, between 1.5 and 2 billion people depend on six big rivers that rise in this part of the world (the Indus, Ganges, Brahmaputra, Yellow, Yangtze and Mekong) for drinking water, irrigation and hydropower. The flow in the Indus, Ganges and Brahmaputra is highly dependent on meltwater from upstream glaciers – and once it's gone, it's gone.

If we succeed in staying below that 1.5°C average temperature increase, it's predicted that around a third of this Himalayan mountain ice will be lost by 2100. At 2°C, it will be two-thirds gone. Anything beyond that, and it's all gone.

CONSEQUENCES FOR THE REST OF THE CENTURY

So, where does all that take us? On a business-as-usual basis, it takes us to a *70-metre* sea-level rise, from Antarctica, Greenland and mountain glaciers combined, over the course of the next few hundred years, depending on complex tipping points about which we currently have very little knowledge. Rising sea levels are going to be a massive problem for humankind for centuries to come. So much for business-as-usual.

Closer to home, the IPCC suggested in 2013 a worst-case scenario of around a 1.1 metre sea-level rise by 2100. In May 2019, however, a paper published in the *Proceedings of the National Academy of Sciences* by scientists at Bristol University in the UK, based on all the latest data about melt rates in Greenland and the Antarctic, suggested a worst-case of a 2-metre sea-level rise by 2100. That's twice the 'official' worst-case scenario of the IPCC.

But could it be even worse than that? Back in 2015, the redoubtable Jim Hansen (who has had a disturbing habit of being well ahead of the game since the mid-1980s), together with sixteen senior climate scientists, suggested we could be seeing a 3-metre sea-level rise – not by 2100 but by 2050. He based this on a rather different kind of analytical methodology: that melt rates in Greenland and Antarctica would not happen in a predictable, linear way, he suggested, but exponentially, doubling every few years. This report was hugely controversial: there's no evidence at the moment of any such exponential effect. But, as we've seen, there's still huge uncertainty in modelling these vast dynamic systems.

Given all that uncertainty, a bit of me is tempted to stick with the painstakingly conservative conclusion of the IPCC's worst case of a 1.1 metre rise by 2100 – until I recall all the new data emerging from both the Antarctic and Greenland since it made its last projection. And then one also has to look at the climate record. As Bill McKibben reminds us in *Falter*: '14,000 years ago, as the Ice Age began to loosen its grip, huge amounts of ice thawed in what researchers call "meltwater pulse 1A", raising the sea level by 60 feet [18 metres]. Nearly 13 feet [4 metres] of that 60 feet may have come in a single century.' Palaeoclimatologists tell us we need to go even further back to the Pliocene epoch between 5.3 and 2.6 million years ago. For much of that time, levels of CO_2 in the atmosphere were pretty much where they are now, average temperatures were 3°C to 4°C higher, there were trees at the South Pole, and the sea level was about 20 metres higher than it is today.

Where does all that leave us?

- Sea-level rise to date is no indicator of sea-level rises to come;

- Melt rates in both Greenland and the Antarctic are already deeply disturbing, and accelerating year on year;
- We should be preparing for a sea-level rise of at least 1 metre by 2100, probably 2 metres, and possibly 3 metres;
- The accelerating loss of ice from the Himalayan glaciers puts at risk the lives of hundreds of millions of people who depend on those river systems;
- We may not know exactly where they are, but there *are* tipping points, and both Greenland and Antarctica *will* tip unless we immediately start cutting back on emissions;
- Halving emissions by 2030 is the only surefire way of minimising the risks associated with melting ice and rising seas.

So many numbers, dates, volumes of ice, sea levels, temperatures, etc. But behind all these data are hundreds of millions of people living in vulnerable coastal areas and cities. Even a 50-centimetre sea-level rise by the end of the century will be devastating. Half the world's mega-cities, and almost 2 billion people, live on the coast. The latest estimates from an organisation called Climate Central, published in *Nature Communications* in October 2019, show that land that is currently home to 300 million people will flood at least once a year by 2050 unless carbon emissions are cut and coastal defences strengthened. This is far above the previous estimate of 80 million. As the lead author of the study, Scott Kulp, said: 'As the tideline rises higher than the ground people call home, nations will increasingly confront questions about whether, how much and how long coastal defences will protect them.' The report suggests that 23 million people will be at risk of flooding in Indonesia by 2050, and specifically refers to the decision taken in 2019 by the Indonesian government to move its capital city from Jakarta to a greenfield site in Kalimantan in Borneo over the course of the next ten years – Jakarta has been subsiding for many years, and is increasingly vulnerable to severe flooding, suffering another major disaster in December 2019. It seems insane, but that's the world we're now living in.

8

FEEDBACK LOOPS AND TIPPING POINTS

'Nature is a totally efficient, self-regenerating system.
If we discover the laws that govern the system, and
live synergistically within them, sustainability will
follow and humankind will be a success.'

BUCKMINSTER FULLER

The year 2019 turned out to be something of a tipping point for a lot of those mainstream scientists who could often be heard over the past few years expressing discomfort at some of their colleagues' more outspoken ways of talking about climate change. But 2019 was different; following a year of one extreme weather event after another, they were forced to acknowledge that things just seemed to be moving so much faster than they had once thought possible. For many of the individual scientists who'd been on the receiving end of their earlier condemnation, that brought little comfort.

But for Professor Peter Wadhams, it must be extraordinarily hard to bear without the occasional outburst of rage. Wadhams is one of the world's pre-eminent polar scientists.

As the director of the Scott Polar Institute from 1987 to 1992, and Professor of Ocean Physics in Cambridge until 2015, he knows exactly what's going on at both ends of the Earth – and if there's one research area for which he's best known, it's the state of sea ice in the Arctic; *A Farewell to Ice* (2015) is one of the most impactful books I've ever read.

This impressive track record provided him with little protection against the climate conservatives in the science community who constantly accused him of overstating his projections as to the speed with which sea ice in the Arctic was disappearing. He was heavily criticised way back in 2007 for concluding his annual assessment with this comment: 'In the end, it will just melt away quite suddenly.' But that's exactly what seems to be happening. Sea ice in the Arctic re-freezes every winter after an annual melting period in the summer. But as each winter period gets a little bit warmer, it sets the stage for an even more extended summer melt. The 2020 summer melt was the second worst on record (satellite records now go back more than 40 years) – second only to the melting in 2012.

As Peter Wadhams would be the first to point out, it's impossible to judge this apparent 'death spiral' simply by looking at any one particular year – it's the trend over time that matters. Every year, however, the US National Oceanic and Atmospheric Administration (NOAA) produces its own Arctic Report Card, and its 2018 report confirmed what had been going on for decades: 95 per cent of the oldest and thickest sea ice (the 'binding agent' that holds the rest of the sea ice together by virtue of its relative permanence) had disappeared over the preceding thirty years. The younger the sea ice is, the thinner it is, and the more vulnerable it is to accelerated melting – even if a lot of it still comes together in the cold winter months.

Even a decade ago, there were extensive regions of the Arctic that still had 'multi-year ice' between five and ten years old. Now there are far fewer of these regions, with very disturbing consequences: just to state the obvious, it takes five years to replenish five-year-old ice that has disappeared – and most scientists believe that this kind of ice formation is *never* going to happen again. Which leads inevitably to the journalists' favourite question: when will the Arctic be ice-free in the summer? Not so long ago, scientists used to offer decadal reassurance – 'sometime in the second half of the century', or vague projections along those lines. These days, it seems that 2040 is now the most optimistic forecast available to us, and there are many scientists who believe it could happen well before then.

> The latest models are basically showing that no matter what emissions scenario we follow, we're going to lose summer sea ice cover before the middle of the century. Even if we keep warming to less than 2°C, it's still enough to lose that summer sea ice in some years.

That was the judgement from Julienne Stroeve of the US National Snow and Ice Data Center in 2020. The principal reason for that takes us back to that all-important 'albedo effect', as explained on page 98. With the Arctic covered in shiny white ice, much of the incoming sunlight was reflected back into space. Dark water absorbs far more of that heat (indeed, almost twice as much as floating sea ice), with the warmer ocean then slowing down the growth of new ice during the winter. The impact of this particular feedback loop on accelerating climate change in general is pretty startling. Experts at the Scripps Institution of Oceanography believe that entirely ice-free summers, should they start to occur every year, could add a massive 0.5°C of warming on top of whatever else is going on at that time. That makes it one massive feedback loop.

A WORLD OF TIPPING POINTS

Which takes us back to the IPCC's Special Report on 1.5°C. If you needed any further reason to be more than a little scared about how badly informed we still are, you need to know that this Special Report, impactful though it undoubtedly is, pays scant attention to feedback loops of this kind. As the Institute for Governance and Sustainable Development put it at the time of its publication in October 2018: 'The report fails to focus on the weakest link in the climate chain: the self-reinforcing feedbacks which, if allowed to continue, will accelerate warming and risk cascading climate tipping points.'

Scientists are now grappling with a lot more of these self-reinforcing feedback loops in many different parts of the world. For instance, if we keep cutting down the rainforest in the Amazon, it could become a net emitter of CO_2 rather than one of our most important carbon sinks. Some 17 per cent of the rainforest has already been entirely destroyed, and an even larger area seriously degraded; as the forest continues to dry out, the tipping point (between sequestering CO_2 and emitting CO_2) could lie in the range of 20 to 40 per cent of forest loss. In the northern hemisphere, warmer temperatures and drier land masses inevitably increase both the risk and the scale of forest fires; as drought takes its toll, there are fewer trees to absorb water and then release it back (through evapotranspiration), which makes things even drier. Large forest fires on the west coast of the USA have become five times more frequent since the 1980s.

In November 2019, an article in the journal *Nature* suggested that we may have already crossed a series of tipping points, or be very close to crossing them, warning that the complex science involved in each of them means that great uncertainty and unpredictability remain. For instance, one of the least anticipated feedback loops exploded into the media between June and August 2019, with extensive wildfires across Alaska, Canada and most particularly in Siberia, where more than 13 million hectares of land were affected. Even Greenland (which is 80 per cent covered in ice) experienced serious wildfires.

This could become really problematic: people often refer to the Arctic north as 'a big tinderbox', with its vast expanses of forest and peatland. Normally, it's simply too cold or too wet to be affected by wildfires – but not in 2019, with temperature records soaring across the entire region. Rising surface temperatures have also been linked to the increase in lightning strikes which ignite the fires. Feedback loops drive feedback loops. Apart from all the CO_2 directly emitted into the atmosphere, these wildfires burn away a lot of the vegetation that helps maintain the permafrost (see below), as well as releasing significant amounts of soot and fine particulate matter that settles on the ice (on land or at sea), creating mini-hotspots which in turn exacerbate the speed of melting.

These are all hugely significant 'positive' feedback loops – where 'positive' has nothing to do with 'good news', but everything to do with things speeding up in the *wrong* direction. As climate scientists have been pointing out to politicians for years, there are very few if any negative feedback loops – i.e. those helping to slow down the acceleration of climate change. Which means that all these positive, self-reinforcing feedback loops bring us closer and closer to some of those dreaded 'tipping points' – defined by the IPCC as '*irreversible* changes in the climate system'. Those are my italics – just to emphasise the weight of that one momentous word: IRREVERSIBLE. The only tipping points explicitly referred to in the IPCC's 1.5°C Special Report are our old friends Greenland and the Western Antarctic Ice Sheet, providing a powerful reminder of just how important it's going to be to do everything in our power to ensure that neither of those massive climate systems is allowed to tip irreversibly.

FEEDING BACK IN THE ARCTIC

As Peter Wadhams puts it: 'Arctic sea ice retreat, directly induced by greenhouse gas warming, has impacts of its own which enhance global change effects on the planet, and will cause disastrous consequences out of all proportion to the original change.' Those impacts include an increase in wave activity during the summer, slowing the growth of new ice in the winter; warmer air over an ice-free Arctic, causing the snowline on land to retreat; as the snowline retreats, the reflectivity of the land surface declines dramatically in early summer, heating up the northern tundra all around the Arctic Circle; warmer runoff waters from snowmelt and Arctic rivers further warm the already ice-free continental shelves ... I'm sure you get the picture.

It's at this point that we have to plumb the growing despair of many Arctic scientists – Peter Wadhams is no longer in a small minority – as they try to assess the impact of all this cumulative warming on the vast areas of permafrost around the Arctic Circle. Nearly a quarter of the northern hemisphere's entire land mass sits on top of permafrost – ground that essentially remains frozen (i.e. at or below 0°C) all the year round, in various combinations of soil, rock or even sand that is held together by ice. On top of that permanently frozen ground sits a much more active layer of soil that thaws during the summer months, and then freezes again during the winter – a terrestrial equivalent of the Arctic sea ice, as it were. That active layer is of course being profoundly impacted by all these feedback effects, which can be pretty dramatic. At outposts in the Canadian Arctic, permafrost is thawing seventy years sooner than predicted. Roads are buckling, houses are sinking. In Siberia, giant craters pock-mark the tundra as temperatures soar. Early this year, one of the fuel tanks at a Russian power plant collapsed and leaked 21,000 tonnes of diesel into nearby waterways.

As the active layer of soil fails to refreeze every winter, the permafrost below is also affected. Permafrost areas have warmed by somewhere between 2°C and 3°C since the 1980s, and as it thaws, all that suddenly unfrozen vegetation ends up emitting huge amounts of both methane and CO_2. And that's wretched news for all of us: the world's permafrost holds twice as much CO_2 as is already up there in the atmosphere.

And that's just the terrestrial permafrost. Off great stretches of coastline in the Arctic Circle, there's a continental shelf, usually between about 50 and 100 metres down – the sea bed here is essentially a seaward extension of the permafrost on land – and trapped within those frozen sediments are billions of tonnes of methane. Until quite recently, these coastal waters were covered by sea ice even in the summer months, ensuring that the water below never really warmed up. But from 2005 onwards, as the sea ice started to recede, the incoming solar radiation started to warm these relatively shallow waters. Which in turn means that water that is now above freezing point is impacting on the Arctic sea bed below – something that hasn't happened for tens of thousands of years. As the sediments are thawed by the warming water above, pure methane is released, and rises to the surface in great plumes of the gas – a process rather poetically described as 'an ebullition'.

Inevitably, there's still a lot of controversy about just how much methane is already being released (it's not easy to measure across such a huge region), let alone how much *might* be released in the future if all the different feedback loops continue to reinforce today's Arctic warming. Peter Wadhams and his colleagues put this to the test, more than five years ago, by modelling what would happen if this gradual warming of the continental shelf process ended up in a huge 'pulse' of methane being released. Their conclusion, if that happened, was that around 50 billion tonnes could be released over a ten-year period. This would amount to an additional 0.6°C of warming before 2040, on top of everything else going on today. It should go without saying by now that this would be completely catastrophic. This is what Peter Wadhams said back in 2016:

> I fear it is a collective failure of nerve by those whose responsibility it is to speak out and advocate action. It seems to be not just climate change deniers who wish to conceal the Arctic methane threat, but also many Arctic scientists, including so-called 'methane experts'. For some such experts, accustomed to only minor Arctic methane seeps, there is some excuse. But they have not woken up to the fact that the environmental conditions now are unprecedented. Some other scientists do clearly grasp what is going on, yet by a psychological process of denial prefer to wish it away.

Here's the point at which the maths becomes almost unthinkable. I explained in Chapter 3 that the cumulative global warming since the start of the Industrial Revolution amounts to around 1°C – give or take. These *potential* additional impacts in the Arctic (0.6°C from melting permafrost and continental ice shelves, on top of the 0.5°C from the melting of the sea ice itself) are therefore **the equivalent of total global warming since the time of the Industrial Revolution**.

Even if we do literally everything we can to reduce emissions of greenhouse gases across every single area of human activity over the next ten to fifteen years (through radical global decarbonisation programmes), these Arctic feedback loops have already developed a deeply disturbing momentum of their own, potentially 'setting at nought' all our collective efforts. Which is why a growing number of scientists are now arguing that we have to intervene *now* before it's too late.

Which inevitably brings the conversation around to that super-controversial idea of geoengineering – in this case, large-scale, human interventions in the Arctic climate system to try to stabilise the sea ice before it's too late. Geoengineering is a *huge* topic, and I've devoted a whole chapter to it later on (Chapter 12), but having emphasised the dire impacts of self-reinforcing, positive feedback loops, it's fascinating to spend a bit of time thinking about the possibility of a wholly man-made *negative* feedback loop designed to slow things down and to avoid what otherwise looks like a full-on apocalypse in the making.

There are lots of ideas swirling around here, most of which are pretty dodgy scientifically, unfeasibly expensive, or far too high-risk in terms of unpredictable consequences – as we know, feedback loops have a way of driving further feedback loops. So, whatever happens, any intervention in the Arctic has to happen on a 'no-regrets' basis (i.e. with *no* downside if it fails), and there's only one idea that seems to meet that criterion.

Thirty years ago, scientists at the Universities of Manchester and Edinburgh in the UK were exploring something called 'the Twomey effect'. This is all about the way in which clouds reflect solar radiation back into space, which appears to depend on the concentration of tiny particles around which cloud droplets form. Their work led to this hypothesis: what if you could make clouds 'super-reflective' by seeding them with tiny droplets of saltwater, thereby reducing the amount of solar radiation reaching the Earth's surface? In 2012, the indomitable Stephen Salter, who even now, in his eighty-third year, continues to seek out inspiring engineering solutions to the world's woes, told the UK's Environmental Audit Committee that he'd already put together plans for remotely operated vessels capable of injecting huge amounts of saltwater into cloud formations during the Arctic summer. Subsequently, he claimed that all he needed would be was around £200 million to carry out preliminary trials and then run an entire fleet of ships in the Arctic over a two-year period. The committee showed little interest.

Does all that sound completely crazy? As you'll see in Chapter 12, I'm as sceptical as any self-respecting environmentalist regarding the extreme geoengineering hinterland populated by opinionated billionaires and self-proclaimed 'climate converts' from the fossil fuel companies. But the situation in the Arctic is *so* dire, and the potential consequences of even ten more years of accelerating warming in that part of the world, with feedback loops looping in more and more destructively, *so* unthinkable, that I've come to the conclusion that a £200 million experiment, with literally no regrets that I can think of, even if it turns out to be a total failure, sounds more and more like a seriously good idea. And, as Stephen Salter pointed out to the Environmental Audit Committee in 2017, £200 million is roughly the price that Paris St-Germain paid for the Brazilian footballer Neymar earlier that year.

I can absolutely guarantee that I will now be on the receiving end of a lot of criticism from climate campaigners for daring to voice support for *any* kind of geoengineering proposal – on the grounds that such advocacy might end up persuading politicians that they no longer have to worry about all the radical decarbonisation measures we now so urgently need. Not only do I believe that to be totally wrong-headed (there's now almost complete consensus that we have to be investing both in mitigation – to reduce greenhouse gas emissions – *and* in adaptation, getting ready for the sort of climate-induced disruption that is already inevitable), but which bit of 'Climate Emergency' are some climate campaigners not hearing? 'We need to do everything. And we need to do it now.' Either we learn to live and thrive within the biophysical limits that constrain all life on Earth, or we don't. It's scary, that's for sure. But it's also as simple as that.

9

OUT OF TIME? JUST IN TIME?

'Delay is the deadliest form of denial'
C. NORTHCOTE PARKINSON

Most of what you've just read in the preceding two chapters, representing today's mainstream scientific consensus, is absolutely not 'common knowledge'. A lot of the politicians swept up in calls to tackle the Climate Emergency very rarely set time aside to get the whole scientific picture in perspective. It's the same with many business leaders, professional bodies, journalists, commentators – in line with the old adage, when it comes to climate change, a little knowledge would appear to go a very long way.

That is now shifting: more and more people are beginning to see just how serious things are, picking up insights into the climate crisis from the news, from various weighty reports, or another protest or demonstration. The wildfires in Australia in early 2020 shocked huge numbers of people all around the world. But we still have to be realistic here: for all sorts of different reasons, the majority of people have yet to engage with the science of climate change in any serious way, let alone to understand the full implications of what we now have to do as a consequence of that scientific consensus.

That makes for a very challenging spectrum for the politicians to deal with. What might be described as the 'climate community' (scientists, activists, some politicians, authors, concerned citizens and the like) probably make up just a few million people around the world. One of the most critical debates within this climate community is whether we've left it all too late to do anything about the situation (i.e. we're already out of time), or whether we can still respond in a timely and proportionate way now that we know how bad things are (i.e. just in time).

That's what this chapter is about. But it remains problematic for the climate community that most people would find such a debate quite mystifying – and possibly even alienating if they feel they're being 'lectured' or blamed for changes in the climate. Whatever happens over the next couple of years, we have to start narrowing that gap by whatever means we can. We simply cannot leave these discussions to a relatively small number of scientists, NGOs and politicians: we need a massive investment in citizen engagement at every level of society.

Broadly speaking, in terms of the out of time/just in time debate, one can point to three different shades of opinion:

1. It's already too late. Full stop.
2. It may well be too late, unless . . .
3. It's definitely not too late.

1. It's already too late. Full stop.

To a certain extent, everybody involved in the debate about climate change agrees that it's already 'too late' for some things if not for others: it's too late, for instance, to get concentrations of CO_2 in the atmosphere back to the level they were at before the Industrial Revolution. This in turn means that it's too late to avoid massive climate-induced disruption over the next few decades. There's more or less complete consensus on that score.

Beyond that, many scientists now feel it's too late to restrict the average temperature increase to below 1.5°C by the end of the century, as we've seen, but still believe it is *not* too late to stay below 2°C. And that means we still have a reasonable chance of avoiding runaway climate change, when things start moving so fast that we turbocharge some of the feedback loops I referred to in Chapter 8, leading to critical tipping points that cannot then be reversed in any realistic timescale.

For me, that's the key question at this time: have we still got enough time to do everything we know we have to do to avoid runaway climate change?

For the hardcore 'too-laters', the answer is a categorical *no*. They look at all the feedback loops that have already been triggered; they point out very reasonably that global emissions have yet to peak, let alone to start coming down; and, finally, they look at the chronic deficits in terms of today's political leadership. For them, taking all that into account, it's definitely too late to avoid runaway climate change. Most of them don't relish finding themselves in that position (it's not easy getting on with a life weighed down by those insights), but they feel very strongly that both the science and political realism are 'on their side'.

This is not a completely new thing. Twenty years ago, I remember listening to author and all-round scientific guru Jim Lovelock talking about 'the Gaia hypothesis' and shocking the large audience in front of him by explaining his own particular brand of 'apocalyptic optimism' – by which he meant that we really didn't have to worry ourselves too much about the future of life on Earth, as life on Earth would recover from runaway climate change and probably be a great deal more resilient on account of the fact that the vast majority of human beings would *not* have survived it!

That is pretty much the same sort of outcome predicted by academic-turned-campaigner Professor Jem Bendell in his 'Deep Adaptation' paper, published back in July 2018. He offers us a disturbing end-of-world continuum, where extreme disruption is seen as completely inevitable, the total collapse of human life and civilisation is seen to be probable, and the outright extinction of the human race a serious possibility. Written initially for his fellow academics (whom he felt were irresponsibly ignoring what the science was *really* telling us), 'Deep Adaptation' helped trigger a timely debate among a much wider constituency of people, with predictably polarised reactions.

Some of those reactions were specific to the timeframe within which his predicted societal collapse will occur. Jem Bendell asserts it will happen before the end of the decade. This is based on a particular hypothesis of his where extreme weather conditions cause the collapse of harvests in many parts of the world, at more or less the same time. This 'multi–breadbasket failure' leads to mass starvation, enforced migration and a breakdown in law and order all around the world – encapsulated in the catchy little acronym INTHE: Inevitable Near-Term Human Extinction! It has to be said that there are not many agricultural experts who go along with him on that hypothesis, even those who are most concerned about the worsening impacts of climate change on food production systems.

Putting that particular dispute to one side, it's not completely clear what Jem Bendell is asking people actually to do, once they've set hope aside and internalised their inevitable grief at the prospect of such sustained trauma for billions of people. There are few 'calls to action' that are likely to have much credibility in such a context, although he does suggest somewhat improbably that 'it might even make our activism more effective, allowing new forms of hope to emerge'. A very similar dilemma confronts the Dark Mountain Project, which ten years ago invited people to give up on fanciful ideas of reversing decades of environmental destruction, and start preparing for a very different 'post-collapse' world. The single most important thing in this world, they argue, will be to protect democracy and human rights in the face of an inevitable upsurge in authoritarian and even totalitarian forces.

In their book, *This Civilisation is Finished: Conversations on the End of Empire – and what Lies Beyond*, activists and authors Rupert Read and Samuel Alexander add real depth to the Deep Adaptation thesis by focusing on some sort of slim hope for a better world even as our current global economy collapses. They see three potential outcomes of climate breakdown: 'this civilisation could collapse utterly and terminally', leading to the extinction or near-extinction of humankind (that's Option 1); 'this civilisation will some-how manage to transform itself radically and rapidly enough to avert collapse' (that's Option 3), which they see as very unlikely; what they're really interested in is Option 2: 'this civilisation will manage to seed a future successor civilisation *as this one collapses*' (my emphasis). As they say: 'Any of these three options will involve a transformation of such extreme magnitude that what emerges will no longer in any mean-ingful sense be *this* civilisation. One way or another, *this* civilisation is finished.'

As I said, people do not arrive at these positions lightly. Such thinking brings with it a complex emotional burden made up of sadness, grief, guilt, anger and fear. In an extremely moving essay, climate scientist Kate Marvel puts it like this:

> There is now no weather we haven't touched, no wil-derness immune from our encroaching pressure ... The world we once knew is never coming back. I have no hope that these changes can be reversed. We are inevitably sending our children to live on an unfamiliar planet. But the opposite of hope is not despair. It is grief. Even while resolving to limit the damage, we can mourn. And here, the sheer scale of the problem provides a perverse comfort: we are in this together. The swiftness of the change, its scale and inevitability, binds us into one, broken hearts trapped together under a warming atmosphere ... We need courage, not hope.

I love the writing, but I find myself at odds with her conclusion. I do *not* believe that courage is an adequate substitute for hope; indeed, there's something disturbingly fatalistic about that idea, conjuring up images of conscripted soldiers 'courageously' marching into battle knowing that they're almost certainly going to die. And I do (still!) have hope that some of the changes we've inflicted on the planet can and will be reversed – which puts my own position much closer to 'It may well be too late, unless . . .'

2. It may well be too late, unless . . .

As I've already mentioned, Jim Hansen, one of America's most respected climate scientists, has spent plenty of time looking at the odds of us avoiding runaway climate change. And those odds look less and less good as the years go by. In June 2018, thirty years on from his ground-breaking evidence to a US Congressional Hearing in 1988, he was excoriating in his condemnation of US politicians, including Barack Obama, for the lack of progress made. But he still professes to have hope: 'It's *not* too late. There is a rate of reduction that's feasible to stay well below 2°C.'

There are now many scientists who feel growing despair at every year that passes without a proportionate response to the crisis we're now in. I referred to the views of Peter Wadhams in the last chapter, who follows an equally outspoken approach here in the UK. And there's no doubt that the work of these scientists (as much as anything from the more 'conservative' IPCC) has had a huge impact on leading authors and commentators – and, more recently, on climate activists. In that regard, it's fascinating to think of this 'out of time/just in time' spectrum through the lens of the climate change campaigners involved in Extinction Rebellion.

Extinction Rebellion (XR)

Both Extinction Rebellion and the school strikes campaign exploded into our lives in 2019, as well as equivalent movements in the USA. They've injected forceful new energy into a climate movement that just felt exhausted. It's impossible to say at this stage whether either of them will still be as influentially active in another five years, or even twenty years, as they are today, but to a certain extent that doesn't really matter. In my opinion, these organisations have already 'changed the rules' in a permanent and hugely significant way.

Two aspects of XR's work merit particular attention. First, its outspoken, consistent emphasis on the importance of science. XR sees this as the best way of getting beyond the highly politicised arguments about who is to blame by urging people to make up their own minds, to review, as dispassionately as possible, what the latest scientific evidence adds up to. Second, and perhaps even more important, is the emphasis on non-violent direct action (NVDA). This is certainly not new. As a member of the Green Party for more than forty-five years, I've always supported the use of NVDA, and am very comfortable that it should be at the heart of the party's philosophical principles. Greenpeace International has led the way, on so many different issues, in many countries, in using NVDA responsibly to highlight particular issues around the world. Local environmental campaigners have used occupations and blockades on countless occasions to slow down and even halt the juggernaut of endless fossil fuel development – most particularly in the USA, as highlighted by Naomi Klein's hugely influential book, *This Changes Everything*. This is a heritage that Extinction Rebellion cares about passionately, drawing on the inspirational legacy of Mahatma Gandhi and Martin Luther King, demonstrating the power of steadfast, principled, radical non-violence.

This theory of change (good science + disruption through NVDA) has been heavily influenced not just by inspirational historical precedents (to which I'll return in Chapter 18), but by the research of Erica Chenoweth in the USA, whose work reveals that any progressive cause needs no more than around 3.5 per cent of the population to be actively involved (that's around 2.4 million in the UK and 11.5 million in the USA), backed up by roughly 50 per cent of the general population talking about the issues in a broadly sympathetic way.

Visiting New Zealand, in November 2019, I was delighted to discover that the 'School Strike for the Climate' in September had been incredibly well-supported throughout the country – the estimated 170,000 young people and supporters on the streets amounted to precisely 3.5 per cent of New Zealand's total population. So New Zealand has already reached its 'tipping point', an impression reinforced just a few weeks later by the passing of its Climate Change Response (Zero Carbon) Amendment Act, making it the tenth OECD economy in the world to commit to a Net Zero economy by 2050. ('Net Zero' means that any emissions of CO_2 and other greenhouse gases are balanced by measures to remove the same amount of CO_2 from the atmosphere – through, for example, reforestation and sequestering CO_2 in the soil.)

For XR, Net Zero by 2050 is far too leisurely a pace; it's called for Net Zero in the UK by 2025, initially demanding that a Citizens' Assembly should be convened to determine how best to achieve that target, and then be given powers to implement the necessary policies. There was a wilful political naivety in this: 2025 is an impossible target date, given where the UK economy is today, and setting out to override our democratic processes (however cumbersome they may be) is a dangerous and foolish approach. But the collective leadership of XR believes that the general public will get behind that kind of approach once people truly understand the nature of the Climate Emergency, and remains deeply frustrated by what it sees as a complete lack of urgency on the part of the mainstream NGOs. One of the first 'occupations' XR organised in 2018 was of the Greenpeace head office in London, which they described as 'a tap on the shoulder of a friend'.

The level of public support that Extinction Rebellion can continue to count on is a moot point. Through most of the summer of 2019, its members were engaged in a fierce debate about whether or not to target Heathrow airport to highlight the growing contribution that aviation is making to climate change. These discussions included the possible use of drones in a rolling campaign all around the airport's perimeter. In my view, 'drone attacks on Heathrow' would be the fastest possible way of destroying much of the goodwill and public support achieved to date.

This widely held view became even more persuasive after XR's protests in London in October 2019 ended in a PR disaster, with an unfortunate fracas involving XR protestors and commuters being prevented from getting to work. Writing in *The Guardian* the day afterwards, Gaby Hinsliff reminded readers that XR wasn't set up to achieve change by winning over public opinion *en masse*, but to force governments to move much faster and much further. However:

These clashes [. . .] feel like an early warning of what can happen when the idea of public consent is brushed aside by people convinced they don't have time for all that. Any political movement refusing to engage with the bread-and-butter concerns of low-paid, marginalised or justifiably anxious people, cannot call itself progressive. That the poorest will suffer most from global heating is no excuse for trampling them in the rush to do something about it.

XR announced a new strategy in February 2020, seeking to build bridges with other campaigners rather than condemn them out of hand for their inadequacies, to be much more creative in taking on the might of the fossil fuel companies, and to make itself more accountable to its members. There's still enormous power in the whole idea of 'rising up in rebellion', and there are now around 650 XR groups in 45 different countries, with countless examples of incredible courage and inventiveness responding to very different circumstances and constraints.

2020 also saw the emergence of different sectoral groupings, including Writers Rebel, Money Rebellion, XR Scientist, and so on. Animal Rebellion has been particularly active, powerfully addressing the nexus between animal right and climate justice – a theme to which I will return in Chapter 13. The essence of its approach, arguing that we may well be 'out of time' unless world leaders *seriously* get their act together over the next decade, still strikes me as entirely reasonable. For all the good things being done by others (as we saw in Chapter 5), it's only governments that can legislate for the kind of accelerated transformation we now need, deploy both 'sticks and carrots' with serious intent to change corporate and individual behaviour, redirect public expenditure, and combine forces internationally to get 'the hard stuff' over the line.

3. It's definitely not too late.

This represents the mainstream consensus position for most of the politicians involved in climate change, for the majority of scientists working through the IPCC, for most of the companies involved in the kind of initiatives I referred to on page 71 and for most 'shiny optimists' (see page 13). This consensus is underpinned by an assumption that 2050 is a 'realistic' target date for countries to be able to achieve a net zero position, but still 'enormously demanding'. And that's probably where most concerned citizens find themselves as well. As we saw in Chapter 5, polls and surveys in almost all countries (including Brazil, Russia, Australia and other countries governed by 'denialists' of one kind or another) reveal surprisingly consistent expressions of concern, but with little awareness (as yet) of the scale and urgency of the transformation required.

As with many people thinking about where they might be on this 'out of time/just in time' spectrum, there's been a definite progression for me personally. When Friends of the Earth started campaigning on climate and deforestation issues back in the 1980s, there was a lot of urgency in what we were calling for, but absolutely no sense of it being any-where near 'too late'. Then came the Earth Summit in 1992, with the all-important Framework Convention on Climate Change as one of its key outcomes, followed by the Kyoto Protocol in 1997. These critical international agreements gave everybody some cause to be hopeful, however frail that hope might have been even at the time. But it was already clear within a few years that world leaders had no serious intention of doing much about it other than to carry on talk-ing. Between then and the Paris Agreement in 2015, against a backdrop of more and more extreme weather events and grim scientific reports, I found it getting harder to hang on to authentic hopefulness.

Right now, in 2020, I'm definitely *not* in the 'too late' camp in terms of us being able to avoid runaway climate change. Indeed, I very much subscribe to the view that it will never really be too late, as there will always be serious campaigning work to be done to prevent things getting even worse in the future than they might otherwise be. As the US climate cam-paigner David Roberts put it back in 2013: 'Remember, there is no "too late" here, no "game over" – it will be a tragedy to shoot past 2°C to 3, but 4 is worse than 3, and 5 is worse than 4. Being unprepared for any of those will be worse than being prepared. The future always forks; there are always better and worse paths ahead. There's always a difference to be made.'

There is indeed. But we need to justify that kind of posi-tion with our eyes open. I myself have no hope, for instance, that we can avoid the horrors of runaway climate change if we continue to believe we can maintain our current way of life, based on today's prevailing growth-at-all-costs economy, with all its built-in wastefulness and disregard for the natural world. What does keep me hopeful, by contrast, is the prospect of a very different kind of 'political tipping point', achieved through an accelerated, radical transformation of political systems the world over, bringing to an end the stranglehold that today's self-serving elites still exercise over the global economy.

As I explained in the introduction, I believe the prospects for that are greater today than ever before – as a direct consequence of the COVID-19 pandemic. The idea of returning to some kind of business-as-usual global economy – continuing to ruthlessly exploit the planet's natural wealth, and to put at risk any prospect of a stable climate – would be sheer madness. And I don't believe people will allow that.

For me personally, therefore, an 'out of time, unless . . .' position means a renewed commitment to more radical campaigning. I believe it's still incontrovertible that our political leaders will not move far enough or fast enough without a great deal of direct pressure being brought to bear on them. Only a sustained period of renewed radical campaigning, including many different kinds of civil disobedience, will provide that kind of pressure – a conclusion to which I will return later.

Part 3

Confronting the Emergency

10

NARRATIVES OF HOPE

'So everyone out there: it is now time for civil disobedience. It is time to rebel.'

GRETA THUNBERG

At the time of going to press, more than 1,900 jurisdictions and municipalities around the world have declared a Climate Emergency, covering around 13 per cent of the global population. Given that only a few hundred climate activists were even talking about a Climate Emergency just eighteen months ago, this has to be seen as one of the most extraordinarily hopeful developments in confronting the reality of climate change. Declaring an emergency is not, of course, the same thing as addressing that emergency through a proper plan of action, let alone implementing that plan of action in a timely and purposeful way. But it is a start – a very welcome start.

In this chapter, I want to look at two very different ways of summoning up a combination of hope and courage in people: the Green New Deal, and, first, the idea of amplifying the urgency behind today's climate crisis by likening it to having to go to war – 'going onto a war footing', as it were, facing such a clear and present collective danger.

GOING ONTO A WAR FOOTING

In wars, people are called on to rethink pretty much everything, to combine forces for the common good, to move fast, to show extraordinary courage and stoicism, to clear the decks of everything that isn't fundamental to the war effort. We need all of this, and more, in confronting the Climate Emergency. But I have always been more than a little nervous about this particular call to action. We may indeed be disturbingly close to a point of no return, as I've tried to lay bare in the preceding few chapters. But using war as the go-to analogy for what we now need to do, given the calamitous impact of armed conflict on humankind, democracy and nature, just feels wrong.

Except when it's a question of taking on the legion of residual climate denialists, inactivists and self-serving elites that stand between us and responding appropriately to the Climate Emergency. I cover this in some detail in Chapter 15, but Michael Mann's excellent book *The New Climate War* reminds us all just how important it is to remain militant in our work against this still powerful force – what he calls the 'architects of misinformation and misdirection'. As he points out, now that outright denial of the science of climate change is no longer available to them, they are now employing every dirty tactic, stratagem and dark art under the sun to delay the transition away from fossil fuels.

We must understand that the forces of denial and delay are using our fear and anxiety against us so we remain like deer in the headlights. I have colleagues who have expressed discomfort in framing our predicament as a 'war'. But, as I tell them, the surest way to lose a war is to refuse to recognise you're in one in the first place. Whether we like it or not, and though clearly not of our own choosing, that's precisely where we find ourselves when it comes to the industry-funded effort to block action on climate.

That makes complete sense to me, and it's what keeps the flame of rebellion burning in so many XR activists and young climate campaigners – whose incredible work in USA and Australia Michael Mann is tirelessly supportive of, and who have had such a hard time from the onset of the pandemic in 2020. Everyone recognises that when young people's campaigns make a comeback (which they surely will), it won't be just with monthly strikes – COVID-19 has made the idea of missing out on more formal education pretty much unacceptable. Instead, young climate activists will be focusing more on local campaigns, developing a diversity of different tactics, with a clear emphasis on a more radical and justice-oriented politics. I see that idea of finding ourselves in a war with the inactivists as being very different from the idea of 'going onto a war footing' in combatting climate change itself.

Moreover, we need to be very aware of the degree to which waging wars today, and preparing for possible wars tomorrow, flies in the face of adopting a low-carbon future in any serious way. Global military expenditure amounted to a massive $2 trillion in 2020, almost 4 per cent up on 2019. That's

$2,000 billion theoretically devoted to 'protecting national security' in one way or another, with the USA responsible for more than a third of this total. We've known for a long time, as confirmed by the Union of Concerned Scientists, that the US military is the biggest single consumer of oil anywhere in the world, with its vast fleets of planes, ships and ground vehicles, but it's important to start thinking about the wider impact that $2,000 billion has on accelerating climate change, not least in terms of the huge opportunity costs that this entails. The IPCC has suggested that an annual investment of $2.4 trillion is needed in the energy system alone, through to 2035, in order to limit the temperature increase at below 1.5°C by the end of the century. That is a huge sum of money for governments and investors to find.

For me, therefore, militaristic imagery in the context of the Climate Emergency is problematic. The idea of 'waging war on carbon' strikes me as particularly nonsensical. We are not 'under attack' from climate change; it's not the CO_2 or the methane that's putting our civilisation at risk, but rather the political system (and the military–industrial complex behind it) that continues to promote aggressive, carbon-intensive economic growth – whatever the consequences. From that perspective, any serious attempt to address today's Climate Emergency must include an explicit condemnation of militarism (and of the arms trade in particular) and a concerted effort to reduce military expenditure, year on year, for the foreseeable future. As well as a peace dividend, it would be one of the biggest climate dividends we could possibly hope for – not least because any country's armed forces will almost certainly be the sector most resistant to full-on decarbonisation strategies.

Climate campaigners very rarely talk about any of this, and it's left mostly to peace organisations, anti-nuclear protestors and campaigners against the arms trade to keep raising these uncomfortable issues. As we beaver away seeking to reduce emissions from power generation, the built environment, shipping and aviation, steel and concrete, land use and farming, you have to ask why we never hear more about emissions from making or preparing for war. This must surely change. Any movement that is enthusiastically advocating (or at least indirectly supporting) the widespread use of non-violent direct action *must* now speak out loudly against the evils of militarism.

But the 'war footing' analogy has many advocates, who love nothing more than a good quote from Winston Churchill. For years, Churchill was dismissed as an alarmist, however hard he tried to point out that he was just raising the alarm – a distinction that seems all too appropriate in today's climate debate. Here he is in 1936: 'Owing to past neglect, in the face of the plainest warnings, we have entered upon a period of danger. The era of procrastination, of half measures, of soothing and baffling expedients, of delays, is coming to its close. In its place we are entering a period of consequences [...] We cannot avoid this period; we are in it now.'

We are indeed. Thirty years of procrastination from politicians the world over has brought us to the very brink of an almost unimaginable disaster for the whole of humankind – a very genuine existential threat. It's precisely our continuing failure to comprehend what that threat looks like that prompts comparisons with the Second World War, and the kind of societal mobilisation that this entailed at the time. The impact of the Second World War on the economies of both the UK and the US was utterly staggering: the UK spent 53 per cent of its GDP every year for six years on that war effort, and the USA 41 per cent of its GDP between 1941 and 1945. We're not yet at the point where addressing the Climate Emergency necessitates such a dramatic reprioritisation of any country's budget – but for how long will that remain the case?

The attack by the Japanese on Pearl Harbor in December 1941 is seen to provide a particularly compelling analogy. For two years, the USA had contributed very little to the Allied war effort, but its speed of response after Pearl Harbor was indeed quite remarkable. Within four days, all car manufacturers were ordered to cease production of civilian vehicles and convert their production lines for military purposes. Other mandates to key industries followed rapidly.

> It's clear that society is capable of responding dramatically to major threats when there is acceptance of a crisis. At that point, all previous arguments against action are consigned to the dustbin. Modern history's strongest example is World War II; others include 9/11 and the recent financial crisis. The evidence of accelerating climate change will build, and then at some point, there will be a 'great awakening' – a tipping point where, relatively suddenly, people will perceive the situation as a real crisis.

That quote comes directly from a rather fascinating paper ('The One Degree War Plan') written by two close colleagues of mine, Professor Jørgen Randers and Paul Gilding, back in 2010. It set out to demonstrate what kind of actions would be needed to reduce greenhouse gas emissions by 50 per cent in five years, and then to start drawing down around 6 billion tonnes of CO_2 from the atmosphere – every year for the rest of the century. The core premise of the paper was that public awareness of the threat to humankind would grow and grow leading up to 2020, so much so that 'the public will demand emergency action to cut global greenhouse gas emissions'. Unfortunately, as we know, that just didn't happen; the best we got, in 2015, was the Paris Agreement, the successful delivery of which will still result in an average temperature increase of more than 3°C by 2100 – rather than the 1°C that Randers and Gilding were talking about in their war plan.

It's into that zone of ever more lethal procrastination that so many new movements, including the school strikes, The Climate Mobilization in the USA, Ende Gelände in Germany, and Extinction Rebellion in the UK, have made such a dramatic entrance. Established back in 2014, The Climate Mobilization's mission is to 'initiate a WWII-scale mobilization to reverse global warming and the mass extinction of species in order to protect humanity and the natural world from climate catastrophe'. That entails persuading cities, states and the federal government to declare a Climate Emergency, and to adopt the organisation's 'victory plan for the Earth' (www.theclimatemobilization.org).

As we've seen, by that one measure of declaring a Climate Emergency, things are undoubtedly moving in the right direction. But all these new organisations face the same question: why is it taking so long for individual citizens to start stepping up in a much more substantive way? For most people, peacetime thinking still prevails. Are we going to have to wait for some kind of climate-induced equivalent of Pearl Harbor – the complete inundation of Miami, perhaps, or the first-ever Arctic summer where the sea ice completely disappears? Or wildfires in California and Australia getting worse and worse every year? It's awful having to summon up the thought of further climate disasters, and it's hard to believe that people might still require an even more extreme sequence of extreme weather events, entailing death and destruction on an even grander scale. But that's where the majority of people would still appear to be.

Finally, any emergency framing around a 'war footing' should not blind us to the continuing disparities in power, resources and responsibilities that exist within and between societies. Even in wartime (*particularly* in wartime), elites continue to thrive at the expense of society as a whole. This is why so many of us balk at every phoney trotting-out of the 'we're all in this together' call to arms, usually by those who would be only too happy to set aside issues of inequality, class, race and gender in their own particular policy response. As Professor John Barry and Dr Noel Healy of Queen's University in Belfast put it: 'Climate breakdown has to be framed as a matter of justice, so that no community is left behind or disproportionately impacted negatively by this transition.' What's more, Naomi Klein's coruscating critique of 'disaster capitalism', highlighting the way in which Western governments and multinationals *always* find ways of profiting from the aftermath of wars and economic shocks, should make us all the more nervous of relying on any kind of war footing analogy. Better by far, surely, to look to a very different kind of precedent in President Roosevelt's 'New Deal' from the 1930s.

The Green New Deal

Back in September 2007, a small group of economists and environmentalists (including the future Green Party MP Caroline Lucas) came together in London to draft a plan to transform the UK economy. They called it 'the Green New Deal'. The full extent of the financial crisis bearing down on us was gradually becoming apparent, in both the UK and the USA, and there were few new ideas around designed to protect both the economy and the environment. It went down extremely well with activists across the progressive political spectrum, but made little impact beyond that, and was soon forgotten as the financial crash brought to a dramatic end what had been a period of relative political and financial stability.

But the Green New Deal group never gave up, and kept on improving and refining a very different kind of recovery strategy for the UK's shattered economy. It became all the more important at that time to establish the link with President Roosevelt's original New Deal, a massively ambitious programme made up of any number of specific initiatives and investments designed to address the ravages of the Great Depression in the 1930s. Equally sweeping reforms were desperately needed in the UK and the US after the financial crash, but the only priority that seemed to matter then was to prop up the banking system at almost any cost. More than $10 trillion was injected into the global economy over the next few years (through quantitative easing and direct bailouts), the principal result of which was to inflate property values and share prices to the huge benefit of the already well-off.

In the UK, the Labour government was swept aside in 2010 as the Lib Dems joined forces with the Conservative Party to usher in what would become nearly a decade of the harshest kind of 'austerity politics'. The Green New Deal went underground, only to re-emerge after the seismic shock of the 2016 EU referendum result and the 2017 general election. At that point, new opportunities opened up for the Green New Deal group to start working with both the Labour Party and the much-chastened Lib Dems; by the time of the 2019 general election, *all* the principal opposition parties (Labour, the Lib Dems, the Greens and the Welsh and Scottish nationalist parties) were prominently highlighting their commitment to a Green New Deal.

For me, this is one of the most remarkable examples of a body of radical but entirely sensible ideas working its way in from the fringes of a political system right to its very heart. But even this uplifting story cannot really compare with the explosion of Green New Deal thinking at the heart of the Democratic Party in the USA – where the resonance of President Roosevelt's 1930s New Deal is obviously that much more powerful. In 2018, Alexandria Ocasio-Cortez was elected to the US Congress as the youngest woman ever. In preparing for her election campaign, her researchers visited the UK to find out about the Green New Deal. Since then, with the enthusiastic support of the Sunrise Movement, she's developed an inspirational appeal around the promise of a Green New Deal to transform the US economy. Together with Senator Ed Markey, a veteran Democrat from Massachusetts, she introduced into Congress a formal resolution for a Green New Deal, with an uncompromising commitment to secure a net zero transformation of the economy by 2030. Most contenders to become the presidential candidate for the Democrats formally backed a Green New Deal in one shape or another, including Senators Biden and Sanders.

President Roosevelt's original New Deal rested on the passage through Congress of fifteen major new bills in his first hundred days in office in 1933, 'from introducing social security and minimum wage laws, to breaking up the banks, to electrifying rural America and building a wave of low-cost housing in cities, to planting more than two billion trees and launching soil protection programmes in regions ravaged by the Dust Bowl', as Naomi Klein puts it. The original proposals for a 21st century Green New Deal in the US had a comparable sweep, based on a series of commitments around renewable energy, sustainable transport, energy efficiency and infrastructure renewal. It guarantees a job for all who want to work, commits to a 'just transition' by promising to protect the wage levels and benefits of all those who lose their jobs in the fossil fuel industries, and rounds out a whole raft of socioeconomic proposals by calling for universal healthcare, childcare and higher education.

What was particularly interesting about the US Green New Deal is that it amounts to a new industrial strategy for America. Once upon a time, the US was one of the most enthusiastic exponents of a top-down, federally driven industrial strategy. In *Concrete Economics*, Stephen Cohen and J. Bradford DeLong state: 'From its very beginning, the United States again and again enacted policies to shift its economy onto a new growth direction – toward a new economic space of opportunity. Yes, there was "an invisible hand", but the invisible hand was repeatedly lifted at the elbow by the government, and re-placed in a new position from where it could go on to perform its magic.' Then came Milton Friedman, Ronald Reagan, liberalisation, and the demonisation of anything to do with 'big government'. The new religion was the power of the market. Markets were going to provide all the answers – and we know how well that's worked out.

The Green New Deal proposals revived the idea of strategic industrial strategy, directing the market through regulation, fiscal policy, public procurement, more effective public–private partnerships, and ambitious public spending. It even sounded a bit Trump-like in places, demanding 'a massive growth in US manufacturing', and 'border adjustments to protect jobs'. The idea of making things again excites US voters. 'This wave of globalization has wiped out our middle class. It doesn't have to be this way [...] We can turn it all around, for our skilled craftsmen, tradespeople and factory workers.' That just happens to be Donald Trump; it could just as well be Alexandria Ocasio-Cortez!

At the time, this was one of the most exciting developments in US politics for many years. It was heart-breaking to see Hillary Clinton lose out to Donald Trump in 2016, but another four years of centrist, corporatist Democratic politics (neo-liberalism with a human face, as it were) – though obviously infinitely better than the horror story that Trump's Presidency proved to be – would *not* have moved the needle very much on either social or climate justice. The revival of Green New Deal thinking represented a definitive break with all that. This is how David Roberts, one of the most perceptive and feisty of US commentators on the climate debate, put it at the time:

> Here's the only way any of this works: You develop a vision of politics that puts ordinary people at the center and gives them a tangible stake in the country's future, a share in its enormous wealth, and a role to play in its greater purpose. Then organize people around that vision and demand it from elected representatives. If [they] don't push for it, make sure they get primaried or defeated. If you want bipartisanship, get it because politicians ... are scared to cross you, not because you led them to the sweet light of reason ... Into this milieu comes a youth movement that takes a Democratic Party disengaged and unambitious on climate change and smacks it on the head. It puts the ultimate goal – to completely decarbonize the US economy in a just and equitable way – on the mainstream Democratic agenda for the first time ever. It accomplishes all this in the course of a few short months.

At the end of April 2021, Joe Biden and Kamala Harris had been in office for nearly a hundred days. From a climate and sustainability point of view, they were one hundred quite extraordinary days, as first flagged on January 27th when Joe Biden rolled out a slew of Executive Orders to start dismantling Trump's poisonous legacy. Bill McKibben described this as 'The most remarkable day in the history of America's official response to the climate crisis':

His orders set up or strengthen offices in the Justice Department, the Energy Department and the Environmental Protection Agency, to focus on what he called 'environmental justice'. He announced that climate change would become a national security priority for the Pentagon. And all of this came after his earlier pledges to re-join the Paris Climate Accord and to cancel the Keystone XL pipeline. There's a shock-and-awe feel to the barrage of actions, and that is the point: taken together, they send a decisive signal about the end of one epoch and the beginning of another. And that signal, most of all, is aimed at investors: fossil fuel, Biden is making clear, is not a safe bet, or even a good bet, for making real money.

There's been no let-up since then. He's put together a powerful climate team, across the whole of his Administration. He's paused new oil and gas leases on federal land, called on the Director of the Office of Management and Budget to identify and then eliminate federal fossil fuel subsidies, and indicated that his Administration would use its very considerable influence with the World Bank and the IMF to reinforce the goals of the Paris Climate Accord. Most significantly, he's also deployed a very powerful tool that President Obama signally failed to do: to use its huge federal procurement budgets to start shifting markets – as with the decision for all federal agencies to start purchasing electric vehicles.

President Biden and John Kerry (his Special Envoy for Climate Change) witnessed at firsthand how the Republicans successfully thwarted Obama's climate strategy, blocking anything and everything he tried to do through the Senate and the courts. It's already clear that a lot of the heavy lifting on climate change will be done via other legislative measures, including the American Rescue Plan Act and the American Jobs Plan that I mentioned in the Introduction.

The hugely ambitious Climate Plan that Joe Biden and Kamala Harris used to such good effect in their election campaign would appear 'to have been taken off the table', and there's much less explicit talk of a Green New Deal. That said, the new CLEAN Future Act (the Climate Leadership and Environmental Action for our Nation's Future) is still waiting in the wings, with calls for 80 per cent of US electricity to come from renewables by 2030 and 100 per cent by 2035.

What lies behind this strategy is a steely determination to undo 40 years of ideological brow-beating from conservatives in the USA aimed at disparaging and diminishing the role of government. Contrast Ronald Reagan's dog-whistle assertion back in the 1980s that 'government is not the solution to our problems: government is the problem', with Joe Biden's passionate defence of the role of government: 'The government isn't some foreign force in a distant capital. It's us. All of us. We the people.'

A lot of this comes down to public support. Twelve years on from Obama arriving in the White House, things now look very different regarding climate change. There are significant majorities of voters who welcome further action, especially when it's tied to new jobs, improved infrastructure and quality of life, and especially when it's tied to an explicit 'levelling up' agenda, seeking restorative justice for the millions of Americans who've suffered disproportionately from environmental pollution. The commitment to ensure that 40 per cent of the benefits of new investment programmes 'will flow to disadvantaged communities' is the Administration's clearest signal yet of its determination to make good on those environmental justice promises (see page 292).

Beyond that, Biden's announcement of a new 'Civilian Climate Corps' on January 27th (putting people to work restoring public lands and waters) came straight out of the Franklin D. Roosevelt playbook – his Civilian Conservation Corps employed 3 million people during the Great Depression in the 1930s. With 19 million US citizens currently unemployed, and young people particularly badly affected, this is an unapologetically 'populist' pitch to the electorate – of a very different kind to Donald Trump's divisive populism.

THE PROBLEM WITH 'GREEN GROWTH'

But just before we get too carried away with all this, we must first confront the potential contradictions at the heart of a Green New Deal thinking: if it turns out to be as dependent on conventional economic growth as our current, very un-green economy, how can that possibly be an answer to the problems we now face?

This is not the place to rehash the whole fifty-year-old debate about economic growth – even though this has been a constant part of my own life as a Green activist ever since I read *The Limits to Growth* back in the early 1970s. But there are basically three, starkly contrasting positions, which can be briefly summarised as follows:

1. *There's really no problem with seeking to improve people's material standard of living through further economic growth,*

as measured by annual increases in GDP. It's served us well over the past seventy years, and will serve us well indefinitely into the future.

I consider this to be the height of economic madness, given the state of the world today, the Climate Emergency, collapsing ecosystems, and continuing grotesque inequality. Eliminating chronic poverty around the world is indeed a burning priority. But as we will read in Chapter 19, carrying on with the same failing, incredibly destructive policies to address poverty cannot possibly be the answer to that massive challenge. So let's not waste any more time on it.

2. *Continuing economic growth on a finite planet is simply not possible; indeed, it's the excessive pursuit of that kind of GDP growth that drives both environmental destruction and social injustice. We urgently need to adopt a zero-growth, 'stable state' economy.*

I get this, and have argued for a long time that all political parties need to prepare electorates for the inevitability of that shift towards a no-growth or very-low-growth economy. But there is little chance of people going along with such a radical reorientation at this stage – especially those who are still in chronic poverty. More than 3 billion people are still living on an income of less than $5.50 a day.

And that's all the more relevant as governments around the world struggle to restore vitality to economies laid low by the impact of the COVID-19 pandemic. I mentioned earlier the IMF's preliminary estimate of the loss of global output amounting to around $28 trillion, and that will get a great deal larger by the end of 2021. That translates into tens of millions of individuals left even less well-off and even more insecure than they were before the pandemic. Concerted efforts to advocate for 'zero growth' at such a time will rightly be dismissed as both unworldly and callous.

3. *Our first priority must be to 'decouple' economic growth from the appalling damage it does to the environment, to the climate, and to people's lives; to transform the way we create wealth, and to get rid of our obsession with GDP, which is a wretchedly inadequate way of measuring economic progress.*

This, in a nutshell, is what 'green growth' is all about, explicitly avoiding the much deeper critique of conventional economic growth (in Position 2 above) by seeking to eliminate its worst excesses and to max-imise its potential socioeconomic benefits. The Green New Deal epitomises that kind of compromise.

This all links back to the previous chapter. One of the strongest reasons for believing that it's already too late to avoid runaway climate change is the apparent impossibility of freeing ourselves from our dependence on conventional economic growth. Most people find it almost impossible to envisage how we might secure better lives for people, let alone how we might eliminate persistent poverty, without that kind of economic growth. Worse yet, there's more and more evi-dence that tells us that 'decoupling' economic growth from the emission of greenhouse gases (Position 3 above) is a great deal harder than might be imagined. A report engagingly entitled 'Decoupling Debunked', published in July 2019, finds no evidence that modern societies have managed to decouple economic growth from emissions at the dramatic scale that is now required.

Take the whole transportation challenge we looked at in Chapter 4. All Green New Deals advocate for an accelerated phasing out of the internal combustion engine, with electric vehicles taking their place just as fast as possible. This does indeed offer a relative decoupling at its best: for every kilometre driven in an EV rather than in a petrol or diesel car, there's a huge reduction in emissions of greenhouse gases – just so long as the electricity comes from renewables rather than fossil fuels. But significant amounts of raw materials have to be mined to manufacture those EVs and the batteries on which they depend, requiring a lot of steel, rubber, lithium, rare earths and so on.

Ideally, we should therefore be seeking a *much* more radical decoupling by dramatically reducing the number of vehicles (of any description) produced every year; by disincentivising private ownership of cars and moving as rapidly as possible to car-sharing schemes, community taxis and first-class public transport; and by prioritising an urban infrastructure that promotes walking and cycling over all other forms of transport. That's still not absolute decoupling (even that kind of integrated, highly sustainable infrastructure entails the emission of some greenhouse gases), but it's an infinitely smarter way of decarbonising the transport system than getting fixated on EVs. As Naomi Klein says:

Every country's Green New Deal must make sure that we actually hit the steep emission reduction targets mandated by science. If we simply assume that by switching to renewables and building energy-efficient housing it will happen on its own, we could end up in the supremely ironic situation of kicking off a Green New Deal emissions spike.

There's real power in these new narratives, combining inclusive and innovative ideas about new economic policy with the first serious initiatives that policymakers have come up with to meet the challenge of radical decarbonisation. That's going to be all the more important in any post-COVID-19 recovery period. If we can maximise the economic upside in doing what we have to do anyway, to ensure a safe and stable climate, then the prospect of building widespread political support around the world becomes that much greater.

As I mentioned in the Introduction, governments' track record in driving those synergies during 2020 was nothing short of negligent, with far more support provided for yesterday's earth-trashing economic sectors than for the 'sunrise industries' of tomorrow. The prospects for 2021, however, are already looking a great deal more positive, driven in part by the Biden/Harris administration, in part by the EU's massive Green Deal, and in part by preparations for the big climate conference at the end of the year.

Encouragingly, the focus is now widening out beyond the relatively 'low-hanging fruit' of renewables and electric vehicles to start getting serious about those 'hard-to-abate' sectors on which all Net Zero trajectories ultimately depend.

THE HARD STUFF

'Hope is not something that you have.
Hope is something that you create with your actions.'
ALEXANDRIA OCASIO-CORTEZ

At the back end of 2018, a high-powered international body called the Energy Transitions Commission issued a report called 'Mission Possible'. The subtitle tells you why this is so important: 'Reaching Net-Zero carbon emissions from harder-to-abate sectors by mid-century' – the sectors in question being cement, steel, HGVs, shipping, aviation and plastics. The commission's overall conclusions are as follows:

- Even with these hard-to-abate sectors, reaching Net Zero emissions is perfectly doable, technically and economically;
- It can be done with minimal cost to the economy and to consumers;
- Significant new innovation is going to be needed, but

a lot depends on driving energy efficiency and managing demand;

- Policymakers (i.e. governments) have got to massively up their game: it may all be doable, but it's not doable without smart regulation and incentivisation;
- Investors should be building up their own expertise in these sectors: as with everything else in today's accelerating transition, trillions of dollars of economic value will be made and lost.

STEEL

Both steel and concrete account for about 8 per cent of global greenhouse gas emissions. The rate of new build (especially in countries like China and India) will not be slowing down anytime soon. Ultimately, we'll have to learn how to construct those new buildings with significantly lower volumes of both. One of the go-to gurus in this area, Professor Julian Allwood of Cambridge University, has mapped out how we can cut the emissions from these two materials by 50 per cent by 2050 simply by designing buildings to use a lot less of them – taking the challenge right to the heart of the architecture profession.

As far as steel is concerned, the first priority is to massively increase the use of recycled steel (on average, new steel products contain 37 per cent of recycled steel), which can reduce emissions by about two-thirds for each tonne manufactured. There's still a problem with the impurities left over in recycled steel, but that's a bigger issue for the car industry than it is for the construction industry. Steel can be recycled pretty much indefinitely.

Once we've got good at using less steel and making sure that more of it is recycled, there's still a huge innovation challenge with the basic production process, both in terms of removing iron from the primary raw material (iron ore), and then in converting that iron into steel in high-temperature furnaces using carbon-intensive coking coal. Companies are constantly trying to drive down costs through greater efficiency (ArcelorMittal's Climate Action Report from May 2019 gives a really interesting account of how much can be achieved in this regard), but ultimately the use of coking coal is going to have to be phased out.

The race is now on to crack that conundrum, especially as the four largest steel producers in the world have all committed to becoming Net Zero companies by 2050. Green hydrogen (see page 168) is the favoured alternative here, with a number of leading companies in Germany, Sweden, Japan and China bringing forward major new investments. As ever, this all comes down to efficiency (Chinese steelmakers cause 2 tonnes of CO_2 to be emitted for every tonne of steel produced, whereas in Europe it is only 1 tonne) and price. Supplies of green hydrogen are both very limited and very expensive. So the price of green hydrogen is going to have to come right down, even as the price of CO_2 is going to have to increase dramatically.

But it will happen. The UK's Committee on Climate Change has recommended that the UK Government should set a 2035 target for Net Zero steel, and the Energy Transition Commission has demonstrated how achieving that target can be done. The government is seriously minded to impose that target – which would put an end to plans for a new coal mine in Cumbria (producing coking coal). This was a more or less absurd proposal anyway, given that the UK is well on the way to becoming the first major industrial economy to get out of coal altogether.

Concrete

If anything, cement is an even tougher challenge. It too depends on a high-temperature process, heating up limestone to 1,500°C in a cement kiln using coal. Many alternatives to coal have been tried, the best being old car tyres which otherwise have to be dumped in landfill sites. Again, hydrogen could be the alternative to the use of fossil fuels. But cement has an additional production problem: as the calcium carbonate in the limestone breaks down, it releases CO_2 – about a tonne of it for every tonne of cement. A huge amount of effort is now going into various innovations to crack this problem, but they are *all* more expensive, which means that no one can afford to make them at scale, which is the only way of bringing the costs down. What's more, people are understandably nervous about alternatives to today's standard concrete products (made up of cement, sand, gravel and water), given that it's hard to test for the kind of safety and durability standards required over many decades.

For the time being, that puts the emphasis back on using concrete more intelligently. So many old buildings get demolished not because they're structurally unsound, but because they're no longer 'in fashion' or can't be repurposed. An important report published in *Material Economics* in 2018, looking at the circular economy in the EU, found that the building sector could shave off as much as a third of its carbon footprint by 2050 *if* buildings were designed to use less material, to last longer, to be more easily repurposed, and if construction waste could be reduced and more concrete recycled.

And that takes us back to one of the key recommendations of the Energy Transitions Commission: governments have got to step up to mandate new efficiency standards for cement plants, new procurement standards (as happened with spectacular success with the London Olympics in 2012), and provide incentives for more innovative processes. Otherwise this is an industry that will remain problematically carbon-intensive, dumping more than 2 billion tonnes of CO_2 into the atmosphere every year – roughly three times as much as the aviation industry globally.

THE FLYING BUG

Most of us take it for granted that we can just jump on a plane and rapidly get pretty much anywhere in the world – if we're lucky enough to be able to afford it. We don't really see it as a privilege any longer; some have even come to see it as a right. So, it's no surprise that demand continues to grow (by around 7 per cent per annum for the five years before the pandemic), and as people's standard of living increases in once-poor countries, there's nothing to say that won't be the case again in the future. Airbus was fond of telling its investors that the number of commercial aircraft in operation would double by 2040 – to 48,000 planes.

Prior to the near-total collapse of the aviation sector as a consequence of the COVID-19 pandemic, aviation's carbon footprint was somewhere between 2.5 and 4 per cent of global greenhouse gas emissions. In future years, that share will just keep on growing as other sectors successfully reduce their emissions over the next thirty years. In the UK today, aviation accounts for around 6 per cent of our national emissions; according to a number of commentators in the aviation sector, if we stay on track to get to Net Zero by 2050, aviation's share will have increased to at least 25 per cent – on a business-as-usual basis. And that's because every other sector will be reducing its emissions.

So, business-as-usual just isn't going to work, especially in any post-COVID-19 recovery period. Either demand has to start coming down (not a particularly popular idea with either the public or politicians), or airlines have to meet that demand with a significantly lower carbon footprint. And that's one hell of a challenge, given how limited their technological options currently are. I've come to understand the full extent of that challenge over the past few years as a result of the long-standing partnership that my organisation, Forum for the Future, has with Air New Zealand. I chair its Sustainability Advisory Panel, flying out to NZ twice a year – something I'm completely transparent about given what that does to my personal carbon footprint. To which I'll return in a moment.

There are basically four things that any airline can do to reduce its emissions today before thinking about alternatives to conventional jet fuel:

- Improve fuel efficiency by at least 1.5 per cent per annum (an industry-wide target), by ensuring it has the most efficient aeroplanes in its fleet, and by minimising the weight carried on each flight;

- Do everything it needs to do on the ground (ground service equipment, using electricity at airport gates, office buildings, etc.) using only 100 per cent renewable electricity;
- Work with other airlines to improve flight management and air traffic control, particularly in terms of optimising take-off and landing glide paths;
- Provide offsetting schemes to encourage both individual and corporate customers to offset their own emissions from flying, as well as offsetting all emissions associated with employees flying for work reasons.

First, let's get to grips with this offsetting business, which for many reasons (some good, some bad) remains controversial. It may be easier just to personalise this. Two flights a year to New Zealand accounts for around 8 tonnes of CO_2. To offset those emissions, Air New Zealand invests in a number of projects including permanent native forestry schemes across New Zealand. In 2019, its FlyNeutral offset scheme purchased more than NZ$2 million of carbon offsets for restoring or establishing permanent native forestry, at a cost of NZ$31 per tonne of CO_2 sequestered. As it happens, Forum for the Future also offsets my emissions, as it does for all employees' international travel, investing in a cookstove project (reducing the use of firewood) in Ghana, purchased through ClimateCare, at the price of £4 a tonne. Two great causes, I'm sure everyone would agree, and a good outcome if one *has* to fly – which Forum for the Future, as an international not-for-profit, cannot completely avoid.

The problem, of course, is that not all offsets produce such

good outcomes, and some offset schemes end up not sequestering or saving much CO_2 at all. And at £4, I would be
the first to agree that even good offsets are still ridiculously
underpriced; realistically, we should be paying at least $50-
plus a tonne to reflect the full cost of carbon in terms of its
contribution to climate change – which is why I personally
choose to pay a premium on my own direct carbon footprint,
over and above those offset payments, via financial support for
a number of environmental organisations.

So there are indeed problems with offsets. But there's going
to be an even bigger problem if airlines can't offset as a result
of NGO campaigning making it unacceptable to the general
public because they see it as some kind of scam. From 2022
onwards (assuming some kind of normality returns to the
sector), airlines will be obliged to offset all the greenhouse
gas emissions caused by any increase in total kilometres
flown since 2020, administered through the International
Civil Aviation Organization's (ICAO) Carbon Offsetting and
Reduction Scheme for International Aviation (CORSIA).
Even before the pandemic, CORSIA was something of a
shambles, and NGOs were up in arms about the way in which
the industry was trying to get away with purchasing the smallest number of offsets while paying the lowest possible price.
Everybody can see that this is a plane crash in the making.

The tragedy is that this could be a massive win–win. Billions
of dollars could be generated through the CORSIA scheme by
the end of the decade, and we desperately need to see investments at this scale in forest restoration (as in the shamefully
underinvested Reducing Emissions through Deforestation and
Degradation (REDD) schemes) and new forestry schemes, as

well as in the restoration of wetlands, peat bogs, mangroves, seagrass and kelp forests. Priced properly, and accounted for rigorously, CORSIA could provide a significant share of that required investment, with the added benefit of this being a long-overdue transfer from the relatively well-off (as in people who can afford to fly in predominantly wealthier countries) to some of the poorest nations on Earth to support forestry and land restoration schemes. And for those who prefer to 'keep it local', it's possible to imagine a completely different kind of offset scheme bringing airlines and farmers' organisations together to help sequester huge amounts of carbon back into our depleted soils.

In the longer term, the industry must of course look beyond offsets. In the past few years, there's been a surge of interest in the idea of electric planes, and several airlines (including Air New Zealand) are now actively exploring the possibility of bringing hybrid jet fuel/electric planes into their short-haul fleets from 2030 onwards. There are now around 200 electric aircraft projects underway internationally, most of them focused on small scale opportunities such as urban air taxis. But Rolls-Royce is pioneering a new generation of electric engines, hoping to beat the current speed record of 338km/h at some point in the near future, using the most energy-dense battery pack ever assembled. Airbus has a prototype hybrid electric plane in development (the E-Fan X), and NASA is developing what it hopes will be the first all-electric plane (the X-57 Maxwell). Boeing and low-cost airline JetBlue are also getting in on the act. But none of this is likely to make any material difference until well after 2030.

Until then, much depends on how quickly various drop-in

substitutes can be developed, either biofuels or Sustainable Aviation Fuels (SAF) as direct substitutes for hydrocarbon-based jet fuel. Several varieties of biofuels have already been certified for commercial use, but they account for a tiny percentage of all aviation fuel used today and remain highly controversial. Once the full life cycle analysis is done, existing biofuels (predominantly ethanol from corn and sugar, or biodiesel from certain vegetable oils, including palm oil) do not actually save that much CO_2 in comparison to the jet fuel they're replacing, divert land away from food production, and can encourage further deforestation. The idea that we should be thinking of a massive expansion of land-based biofuels to substitute for jet fuel is now considered wholly inappropriate.

Which is why the focus is now very much on Sustainable Aviation Fuels – essentially turning processed household waste, agricultural wastes or industrial emissions into jet fuel substitutes. There is a huge amount of innovative work going on around the world to develop new fuels on a commercially viable basis. Here in the UK, for instance, ESSAR Oil (which currently processes more than 16 per cent of the UK's transport fuels) has launched new Joint Venture with Fulcrum BioEnergy to build a major new plant near Liverpool, turning non-recyclable household waste into a synthetic jet fuel – with a 70 per cent reduction in CO_2 emissions.

One of the real pioneers in this space is a company called LanzaTech. For the past fifteen years, it's been exploring the use of naturally occurring bacteria to produce fuels, chemicals, plastics, synthetic rubber and suchlike, feeding those bacteria with 'recycled CO_2' from industrial plants or other carbon- based gases from municipal solid waste, industrial or agricultural waste. At its plant in China, emissions of carbon-rich gases from a huge steel plant are piped into a number of fermenters stuffed with bacteria, producing 16 million gallons of ethanol a year. Similar plants are under construction in Europe, India, South Africa and the USA.

LanzaTech has also produced a drop in jet fuel and has successfully flown commercial flights with Virgin Atlantic and the Japanese airline ANA. This jet fuel technology is currently being scaled to 10 million gallons per year of production capacity. British Airways has recently announced plans to invest in a major new plant in the USA, in partnership with LanzaJet (a spin-off of LanzaTech) to convert ethanol from agricultural waste into commercial-scale jet fuel.

But all such alternatives are currently two to three times more expensive than conventional jet fuel, and it's clear that this will become an even more competitive space in the future – including competition with algae-based products.

Ever since I visited a plant in Brazil run by a joint venture called Solazyme Bunge Renewable Oils, I've been a big fan of algae. Solazyme was providing the algae, and Bunge the sugar to feed them up, and it really was a bit like magic! The algae are put into a soupy mix of sugar and nitrogen where they multiply rapidly, and are then transferred into one of six huge fermenters where they go on getting fatter and fatter. At a critical moment, no more nitrogen is injected into the mix, which means they stop growing and dividing, and start converting all that sugar into oil over several days – at which point everything is transferred to a great big dryer to evaporate the water, leaving either an oil-based product or a golden power.

This business hadn't just popped up in a great big Brazilian sugar plantation out of nowhere. Forty years ago, the US Department of Energy set up its Aquatic Species Program to test alternatives to crude oil — an obvious area of inquiry given that the oil we use today is made up of countless trillions of dead algae crushed underground over millions of years. As Ruth Kassinger put it in her intriguing book, *Bloom: From Food to Fuel, The Epic Story of How Algae Can Save Our World*: 'Why not speed up the process and make oil from algae living today? Instead of pumping fossil oil out of the ground and adding carbon dioxide *to* the atmosphere, we'd be growing algae and pulling carbon dioxide *from* the atmosphere.'

There are basically two ways of doing this: in outdoor ponds or water channels, using sunshine as the source of energy; or indoors in great big fermenters like the ones I visited in Brazil, using sugar as the basic feedstock. There are advantages and disadvantages to each method, but both work. Over a couple of decades, a number of different companies were battling it out through the Aquatic Species Program to see which worked best. Both Solazyme and a company called Sapphire, which had the world's largest outdoor algae farms, succeeded in making what became known as 'green crude' at some significant scale. In 2012, the US Navy put on a pretty impressive demonstration (dubbed 'the Great Green Fleet') to fuel a strike force of destroyers, a cruiser and a lot of jets and helicopters with a hybrid fuel made up of 50 per cent conventional jet fuel and 50 per cent algal oil (from both Sapphire and Solazyme) and used cooking oil! Everything worked perfectly; nothing fell out of the sky.

But even as these two companies (and many others) were competing with each other, their real contest was with fossil fuels. When the price of oil was high, they could go head-to-head without any problem; as soon as the price of oil came back down, they just couldn't compete. With oil at $120 a barrel in 2011, it was a bit of an algal bonanza; when it came right down to $30 a barrel in 2015, they were scuppered. Sapphire went out of business in 2017, and Solazyme (now called TerraVia) hung in there through a series of financial crises until it was acquired by a big Dutch chemicals company called Corbion in the same year. All of which means that we shouldn't give up on algal oil as the go-to biofuel substitute for the aviation industry at some stage in the future. It's technologically proven (ask the US Navy), scalable anywhere in the world that can grow sugar, with minimal impact on the world's oceans, forests or farmland. All it needs is a proper price on carbon – just as is the case with so many other potential solutions.

Governments can help drive this, but there's a lot more they're going to have to do to reduce aviation's carbon footprint: prohibit any new airport expansion (including the cancellation of the near-insane proposal for a new runway at Heathrow), and implement effective, equitable ways of ensuring that those of us who fly pay the necessary price. To my mind, that can only be done with what is called a Frequent Flyer Levy: the more you fly, the more you pay. And *all* airlines need to get on the same page (supporting policies like this, as well as rigorous offsetting standards, proper carbon pricing, taxing jet fuel – which is currently exempt from the kind of duties other forms of transport have to pay – and getting rid of their air-miles schemes) before the so-called '*flygskam*' – flight-shaming – phenomenon (people starting to feel guilty when they get on a plane) becomes a much more serious threat to their business models.

According to Swedavia, the company that looks after ten of Sweden's biggest airports, the number of passengers passing through its terminals (pre-COVID-19) fell by around 4 per cent in the first half of 2019. At the same time, the country's state-run rail operator, SJ, reported an almost 10 per cent increase in train journeys during the same period. Similar trends have emerged in Germany and the Netherlands. Virgin Trains in the UK has claimed that record numbers of people chose rail travel over flying between London and Scotland in 2019, a trend that may well be reinforced by a campaign called 'Flight Free UK', which aims to persuade 100,000 people to give up flying, and is busy campaigning for a tobacco-style ban on airline advertising.

It will take a long time for the aviation sector to recover from the COVID-19 pandemic, and it's crucial that industry leaders plan for that with their eyes fixed firmly on the ultra-low-carbon world that awaits us.

Green hydrogen

In September 2020, Airbus raised the stakes in the world of sustainable aviation when it launched concept designs for three new 'zero emission' commercial aircraft to enter service by 2035. The ZEROe initiative is based entirely around the use of hydrogen to replace conventional jet fuel. In the words of Guillaume Faury, Airbus CEO:

This is a historic moment for the commercial aviation sector as a whole. I strongly believe that the use of hydrogen – both in synthetic fuels and as a primary source for commercial aircraft – has the potential to significantly reduce aviation's climate impact. The transition to hydrogen as the primary power source for these concept planes will require decisive action from the entire ecosystem.

You can say that again, Guillaume! But this is just one of many over-the-top aspirations out there in the now massively hyped world of green hydrogen. In just a couple of years, green hydrogen has gone from 'interesting innovation niche' to the apparent answer to all our decarbonisation challenges in the hard-to-abate sectors. Caution is advised in appraising these multiple aspirations.

Hydrogen as we know it today is the very opposite of a green fuel – despite the UK government's efforts to badge it as a 'clean energy technology' in its 2020 Energy White Paper. 97 per cent of the 115 million tonnes used globally (primarily in refining and chemicals) is either 'brown hydrogen' (from the gasification of coal) or 'grey hydrogen' (from natural gas), between them emitting around 830 million tonnes of CO_2 – 2 per cent of total global greenhouse gas emissions.

As to the remaining 3 per cent, there's a tiny amount of so-called 'blue hydrogen' – essentially grey hydrogen but with its CO_2 emissions captured and stored – with the rest made up of 'green hydrogen' from electrolysing water, both of which are much more expensive than the climate-wrecking brown and grey hydrogen.

The gulf between that current reality (rarely mentioned

by hydrogen enthusiasts) and the prospect of readily available and affordable green hydrogen is absolutely vast. That has to be borne in mind even as we hear more and more of efforts to bridge that gulf, including the Hydrogen Council (which anticipates investments of around $300 billion over the next decade) and the newly-launched, UN-backed Green Hydrogen Catapult (involving a number of big companies planning to halve production costs while massively increasing global production up to 25GW by 2026). In the UK, the 'Hydrogen Strategy Now' consortium was set up on November 2020, with plans to pump £3 billion into a UK-wide hydrogen economy.

Both the French and German governments have committed around €7 billion each to new hydrogen programmes. Saudi Arabia is building a $5 billion plant powered entirely by solar and wind energy to be in operation by 2025, with the intention of becoming the world's largest supplier of green hydrogen, driving costs down from around $5 per kg to around $1.50 by 2030.

Somewhat improbably, however, it's Australia that currently heads up the ambition stakes. It has plans to build a massive $53 billion plant in the northern part of Western Australia, called the Asian Renewable Energy Hub, with arrays of electrolysers powered by 1600 giant wind turbines and 78 sq km of solar panels.

Australia eagerly anticipates huge markets for green hydrogen opening up in Japan, South Korea and China, as their decarbonisation challenges become more acute. Powering industrial economies with its sunshine and wind rather than its coal and iron ore has to be Australia's best bet in a world transformed in this way.

Cost will still be a critical factor, and carbon pricing will still be essential if green hydrogen is to compete on a less unequal playing field. Even then, green hydrogen will always be a premium energy source and will never be able to compete directly with the low-cost renewables on which its manufacture depends. This makes pipedreams about substituting hydrogen for conventional gas in the UK's gas grid or producing millions of tonnes of blue hydrogen using Carbon Capture and Storage look entirely insane. Decarbonising steel production, the manufacture of cement, and shipping (with an eye to aviation in the future) should be the priorities for green hydrogen.

DRIVING INNOVATION

Aviation and the built environment (through its use of steel and concrete) represent two of the hardest sectors to sort out. But it's all perfectly doable – as long as we rapidly ramp up investment in innovation. For instance, in researching this book I've learned that some of the chemistry around CO_2 itself is still relatively undeveloped, including how to break the bonds between the oxygen atoms and the carbon. We need to be thinking much more strategically about recycling CO_2 to prevent it getting into the atmosphere in the first place. This means putting the emphasis more on carbon capture and *use* (CCU) rather than carbon capture and *storage* (CCS), including opportunities to use CO_2 to make synthetic fuels. In June 2018, an article in the energy journal *Joule* showed that a company called Carbon Engineering had successfully taken CO_2 out of the air (at around $100 a tonne) and combined it with hydrogen (electrolysed from water using renewable electricity) to make a liquid fuel. 'Air to Fuels' (A2F) now has a number of big backers, including Bill Gates, and although the price of the resulting fuels is still far higher than conventional hydrocarbon fuels, that could all change if we were paying a proper price for carbon.

And nowhere is innovation needed more than in the world of plastics, which has been a topic of huge public concern since the BBC's *Blue Planet II* documentary in 2017 highlighted the risks of discarded plastic packaging getting into our oceans. It's good that both governments and businesses (particularly those that use huge amounts of plastic packaging) are now focused on addressing the 'ocean plastics' issues. What's not so good is that a lot of bad decisions are being made to get rid of plastic even when that's not necessarily the right thing to do, both from an environmental and an energy point of view.

Many of the problems don't actually lie with plastic itself, but rather with the lack of proper systems for managing plastic properly to keep it out of the environment. For example, lots of retailers are now getting out of plastic bags and into paper bags – despite the fact that every single life-cycle analysis ever done in this area shows that this will be worse for the environment in terms of water consumption, water pollution, greenhouse gas emissions and waste. Paper scores much better in terms of avoiding litter, but on nothing else. The simplest (but most controversial) thing would be for governments everywhere to ban the sale of *all* disposable bags; in countries that have done this (including many in Africa), people quickly get used to the idea of bringing their own.

Both in the UK and the US, we're facing a particularly

problematic situation. Manufacturers can use pretty much any combination of plastic they like for packaging purposes, which makes recycling much harder. A lot of experts believe that the choice should be limited to PET, polypropylene and high-density polyethylene, which would massively improve levels of post-consumer recycling. Local authorities in the UK can use pretty much any collection system they like (most countries in Europe mandate one system to be used across the entire country); and most people remain massively confused about what can and cannot be recycled.

Carbon pricing

There's no shortage of ideas out there, and, even more encouragingly, no shortage of capital to invest in good ideas. What we are short of is economically literate politicians who understand just how critical it is to sort out this chronic and life-threatening market failure in the way we manage carbon. There have been many points in Chapter 4 and in this chapter where I've suggested that effective policymaking depends on 'getting a proper price on carbon'. As explained on the Carbon Tax Center's excellent website, that can be done either through carbon trading schemes (like the EU's Emissions Trading System, which allows countries to purchase carbon credits through an open marketplace), or carbon taxes (twenty-eight countries or regions already use carbon taxes, levied on all fossil fuels, and usually paid by the fossil fuel companies themselves). New tax schemes of one kind or another are coming on stream all the time – so we're moving in the right direction, but nothing like fast enough.

Without proper carbon pricing, governments, businesses and individuals will go on happily using fossil fuels, which benefit from a massive subsidy by not having to internalise the cost of all the damage done through the emission of greenhouse gases. In 2017, the High-Level Commission on Carbon Prices estimated that achieving the goals of the Paris Agreement would require a carbon price of between $40 and $80 for every tonne of CO_2 by 2025, rising to $50–100 a tonne by 2030.

The UK has had a 'carbon floor price' since 2013, which functions as the minimum price that fossil fuel companies pay to emit CO_2 if the price of carbon through the EU's emissions trading system falls below that threshold – which it has done in almost every year since 2013. CO_2 emissions in the UK have fallen steadily during that time, not just because of the floor price, but because of lower prices for both gas and renewables. The decision taken in 2015 to phase out *all* remaining coal-fired generation by 2025 effectively killed off any new investment in coal; in April 2021, there were just three coal-fired power stations still operating in the UK. No other country has done as well as the UK in an accelerated phase-out programme of this kind. Carbon pricing works!

But carbon taxes are invariably controversial. The carbon tax in British Columbia became a source of intense political debate in the Canadian general election in 2019, with predictably polarised positions on whether the province's ten-year experiment with its carbon tax had or hadn't worked. The answer is both: if you take into account that the province's emissions of greenhouse gases were exactly the same in 2017 as they were in 2007, you would be inclined to say it hadn't worked. But when you take into account that per capita emissions had been cut by 12 per cent, and emissions per unit of economic output by nearly 50 per cent, against a backdrop of both population growth and economic growth, you'd be inclined to say that it had worked.

Canada also has a federal carbon tax (introduced via the Greenhouse Gas Pollution Pricing Act in 2018), which was initially challenged by oil-rich Alberta and other provinces, but now confirmed as 'constitutional' by Canada's Supreme Court. The tax is levied at CA$30 a ton, with plans for it to increase to CA$170 by 2030. The tax is revenue neutral, with the proceeds distributed as a quarterly 'carbon dividend' so that people in the bottom two-thirds of income levels in Canada get back more than they had to pay in the first place.

Any intervention to price carbon has to be handled with great care. One of my co-founders of Forum for the Future is the eminent environmental economist Paul Ekins. He has argued passionately throughout his life that you cannot increase the price of fossil fuels by imposing a carbon tax if the people who are hit hardest by it are already among the poorest in society.

That's why there's now renewed interest in the USA (particularly among those advocating for a Green New Deal) in the idea of a redistributive carbon tax, ensuring that the burden of the tax falls on those who really *can* afford it, rather than on those who can't, and then redistributing (all or part of) the resulting revenues as a dividend to all US citizens, with exactly the same amount paid to each and every citizen. There are now a growing number of eminent US economists (including the past four chairmen of the Federal Reserve, fifteen former leaders of the White House Council of Economic Advisers and twenty-seven Nobel laureates) recommending exactly that kind of policy initiative.

Weirdly, however, there are now a number of environmentalists in the US opposed to carbon pricing on the grounds that it might jeopardise all the class actions now under way to hold oil and gas companies to account for historical emissions, or that it represents a misguided 'market mechanism' when what is really required is clear Federal legislation. This seems to me to be completely crazy. I can understand why many environmentalists are so nervous about the idea of 'putting a price on biodiversity', but on carbon?!

The likelihood of such a proposal making much headway even under the current US administration is of course remote. And that's deeply worrying: if we can't get global alignment on the absolute basics of addressing today's Climate Emergency through the right economic and fiscal policies, then the decade left to us to halve emissions of greenhouse gases will just fade away. As that impasse persists, the debate about geoengineering is just going to get hotter and hotter – as is the climate.

12

WE HAVE TO TALK
ABOUT GEOENGINEERING

'Hope is something we do rather than have.'
JOANNA MACY

Having done everything we can to stop turning up the heat, we have to start, right now, *turning down the thermostat*. What this means in practice is getting as much as possible of the CO_2 that we've been putting into the atmosphere over the past fifty years back out of the atmosphere just as soon as we can. This is usually referred to as greenhouse gas removal, or (somewhat more geekily!) negative emissions techniques (NETs). And the even bigger, hugely controversial idea behind all of this is, of course, GEOENGINEERING.

The Royal Society's definition of geoengineering is pretty simple: 'The deliberate large-scale manipulation of an environmental process that affects the Earth's climate in an attempt to counteract the effects of global warming.' 'Deliberate' is an important word here. It could be argued that we've been 'geoengineering' our climate – albeit not deliberately – from that very first moment when our ancestors started burning coal millennia ago, making it possible to do so much more than we could with just firewood or other kinds of biomass at our disposal.

Fast-forward thousands of years to the middle of the eighteenth century, when we learned how to use that coal to produce steam to drive engines to power an industrial revolution, through to the end of the nineteenth century, when major discoveries in Texas set the stage for the current oil economy, and you can easily begin to appreciate the momentum that built up around the extraction and conversion of fossil fuels over the past 250 years. In one way or another, we're all here today, all 7.8 billion of us, because we've proved to be such successful and applied geo-engineers.

However, we did all that pretty much unknowingly, and certainly not with any deliberate intent to influence the climate. Indeed, it was pretty much 'unthinkable' to imagine that the CO_2 released from burning all that coal, oil and gas could possibly make any difference − with CO_2 making up no more than 0.04 per cent of all the gases in the atmosphere. Our largely unknowing state persisted well into the 1970s, when the scientists at ExxonMobil and other big oil and gas companies worked out what the consequences would be of decades of breakneck fossil fuel development. By releasing more greenhouse gases (from the burning of fossil fuels and deforestation) than living systems on land and in the oceans can possibly absorb, we've comprehensively destabilised the climate.

In other words, like it or not, we humans now have no option but to try to get the climate under control again.

Having geo-engineered ourselves into this mess, we're going to have to geo-engineer our way out of it, mindfully and responsibly. And very urgently indeed.

NATURAL CLIMATE SOLUTIONS
(OR NATURE-BASED SOLUTIONS)

But there's geoengineering and there's geoengineering, including what might be described as 'bio-geoengineering', or natural climate solutions. And there's no better place to start than planting trees – billions and billions of them. A report published in *Science* in July 2019 estimated that a worldwide tree-planting programme on more than 1.5 billion hectares of land that currently has no trees on it, and is not being used for crops or human habitation (that's about 11 per cent of all available land, equivalent to the size of China and the US combined), could remove around two-thirds of all the emissions that have been pumped into the atmosphere since the start of the Industrial Revolution. Even if the estimate is out by a factor of ten (and it has to be said that this article has come in for a lot of criticism since publication), this is indeed 'mind-blowing', as one of the report's authors commented.

One of the most important conclusions in the report is the cost of getting all that CO_2 sequestered in that way – estimated at around \$300 billion to get the 1 trillion trees planted that would be needed to achieve this mega-sequestration of CO_2. This makes it by far the most cost-effective solution that has ever been proposed for helping to turn down the thermostat.

However, we will need to develop this opportunity very cautiously; there are inevitably some significant problems associated with tree planting on this scale, and the original study has been fiercely criticised by a large number of academics. There could be some seriously damaging impacts on biodiversity if the trees are planted in the wrong way, or in the wrong places, particularly on natural grasslands and savannahs. And it would take between fifty and 100 years – so it's certainly not a 'quick fix'. Beyond that, many of the areas singled out for reforestation are currently grazed by livestock; without freeing up hundreds of millions of hectares currently used to produce meat and dairy, the ambition is completely unrealisable. But that's the point of thinking about this from an emergency perspective. Either we continue to use those hectares to produce meat and dairy products in ways that are dramatically accelerating climate change (as we'll see in the next chapter), or we use them to help remove greenhouse gases from the atmosphere, and in so doing significantly enhance our chances of staying below 2°C by the end of the century.

Whatever the ins and outs of the true potential of this, we should certainly be doing everything in our power to encourage drought-prone countries like Ethiopia (with an ambition to plant 4 billion trees), Pakistan (Prime Minister Imran Khan launched Plant for Pakistan back in 2018, with plans to plant 5 billion trees by 2023) and China, which really leads the world with its Great Green Wall project, planting around 66 billion trees in the north of the country since 1978 – although with somewhat mixed results in terms of survival rates.

Beyond tree planting, we need to be investing now in other natural climate solutions. That includes some of the huge opportunities available to us through restoration of wetlands and peatlands (vegetation sinks to the bottom in still water, locking up carbon as peat), further research on rotational grazing (moving cattle on quickly from one block to the next) and different agronomic systems to help build up soil carbon (as I'll explore in Chapter 17), as well as everything to do with so-called 'blue carbon' – looking at restoring mangrove swamps, salt marsh and seagrass beds, kelp, encouraging fast-growing algae in open ponds and the like. Professor Tim Flannery of the Australian Museum in Sydney, for instance, is a passionate advocate for growing seaweed in the open ocean, pumping up nutrient-rich deep water to the surface. The seaweed then uses these nutrients to trap huge amounts of CO_2 before sinking back down into the ocean.

Closer to home, there's a new focus on the seagrass meadows that once surrounded the nation – more than 90 per cent of which have been lost over the course of the last hundred years as a consequence of pollution, dredging, bottom-trawling and industrial development. There are now only 8,500 hectares left. Seagrasses provide an amazing habitat, storing carbon 35 times faster than tropical forests, harbouring up to 40 times more marine life than bare seabeds, while providing effective protection against coastal erosion by helping to absorb the impact of storms. In March 2021, WWF teamed up with the brewer Carlsberg to help raise funds to protect this critical habitat – with 50p coming from every special edition Carlsberg pack that is sold – as part of its Together Towards Zero strategy.

By some calculations, about 10 billion tonnes of CO_2 could be removed from the atmosphere every year with some or all of these natural climate solutions – that's about a fifth of our current annual emissions, at prices per tonne sequestered that fall anywhere between \$15 and \$50. So just stop a moment to ask why it's so incredibly rare to hear any politician extolling any of these natural climate solutions apart from the usual tree-planting rhetoric? The closest they get to enthusing about anything else 'natural' is through their support for a technology called BECCS – bioenergy with carbon capture and storage. The Intergovernmental Panel on Climate Change itself is almost ludicrously keen on BECCS, suggesting that it's pretty much the only way of staying below that 2°C threshold by the end of the century.

Before we grapple with BECCS, a quick word about Carbon Capture and Storage itself (CCS). CCS has been part of the suite of geoengineering 'solutions' for almost as long as I can remember, the preferred technological darling of the very same oil and gas companies that spent hundreds of millions of dollars to obscure the science of climate change, but could never quite stump up the cash to have a proper crack at CCS. At its simplest, it entails retrofitting existing gas- and coal-fired power stations (and other carbon-intensive facilities) with complex technology to capture up to 90 per cent of the CO_2 that would otherwise be emitted into the atmosphere. It's expensive, and it's also very energy-intensive as it takes a lot of energy to extract and compress a tonne of CO_2 in this way. Once captured, the CO_2 has to be 'sequestered' – disposed of under the ground in geological formations from which (it is hoped) it will never escape.

As of early 2021, there were twenty-seven functioning CCS facilities on fossil fuel power stations around the world, most of them part paid for by using the sequestered CO_2 to inject into existing oil assets to 'enhance recovery' – i.e. get more oil out than would be possible without the reinjected CO_2. That in itself is something of a problem. We need less oil and gas coming to market, not more. We need fewer fossil fuel facilities, not more. As of now, the going rate for a tonne captured and sequestered via this carbon-intensive and still immature technology is anywhere between \$70 and \$120, depending on different conditions.

So what of BECCS – the bioenergy version of CCS? This would entail planting millions of hectares of fast-growing trees (one proposal suggests an area three times the size of India would be required), burning them as biomass in power stations to produce electricity, capturing the CO_2 by the selfsame clunky and carbon-intensive system, and then burying that captured carbon underground. As the IPCC doggedly continues to point out, the potential for BECCS could be significant, but the reality is that no one has shown it can be done affordably on a large scale – and there are as yet no major schemes up and running.

The huge biomass power station at Drax (which has an ambition to become a Net Zero company by 2030) has been trialling a number of small BECCS schemes since 2019. In early 2021, it applied for a Development Consent Order for a major commercial scale BECCS plant, with the expectation that it will eventually be able to capture the equivalent of 8 million tonnes a year.

Many would argue that it still makes sense to position BECCS at the 'potentially viable but still ill-advised' (for the reasons given above) end of the geoengineering spectrum. I've had a go at my own personal take on what this geoengineering spectrum looks like, in terms of removing CO_2 from the atmosphere, at the end of the chapter. Elsewhere on that spectrum, there are two big ideas for removing greenhouse gases from the atmosphere that dominate today's debate over and above the natural climate solutions referred to above.

First, there's ocean fertilisation: spreading iron filings over the surface of those parts of the ocean that happen to be short of iron — a deficiency that curtails the growth of plankton. The idea is that if we provide the missing iron, the plankton will promptly bloom by sucking up CO_2 from the air, briefly thrive and then die, sinking to the bottom of the ocean with all that CO_2 on board. It's relatively cheap, easy to administer, and could be up there in the gigatonne range — as in capable of sucking billions of tonnes of CO_2 out of the atmosphere on an ongoing basis.

Too good to be true? Almost certainly. Not only have we no way of predicting what the impact of such a dramatic disruption might be on sensitive ocean ecosystems, but it's very unlikely that more than a small percentage of those CO_2-enriched plankton would get anywhere near the ocean floor, being gobbled up by crustaceans, fish and other marine organisms all the way down. Which means all that CO_2 would stay in the system.

Next, there's what's called 'direct air capture' (DAC), stripping CO_2 directly from the air via a variety of different chemical processes, powered by renewable energy, and then either liquefying it so that it can be pumped underground or converting it into other materials — as I touched on in Chapter 11.

As it happens, I included a chapter in *The World We Made* on DAC, not because I personally believe it's a particularly smart way of addressing our Climate Emergency, but because I fear it's almost inevitable that we'll end up doing it anyway, at huge expense, simply because we won't have done enough to make such a costly, clunky technology unnecessary. I got a lot of stick from some of my more radical colleagues for even giving voice to such heresy, but the reality is that there are a lot of very smart people working in this area, backed by a lot of reasonably clean money. It's not doing anyone else any harm, so my view is that we should see 'the best they've got', as it were. Costs will almost certainly come down; technology barriers will be cracked. So, let's keep trialling the different varieties of DAC technology.

And let's welcome the latest big play from Tesla-entrepreneur Elon Musk to create a $100m prize (through the XPrize Foundation) for the best carbon removal technology, including Direct Air Capture as well as all the nature-based solutions I've already referred to. We're going to need a lot more of that kind of innovation incentivisation.

Solar radiation management

We need now to take a look at the alternative geoengineering portfolio, which is not about removing CO_2 from the atmosphere, but about seeking to limit incoming solar radiation. 'Solar radiation management' (SRM) is what this alternative portfolio is called, and there's a lot of money going into this as well, including *big* money from the fossil fuel industries.

The go-to favourite here is to create our very own 'shield' in the stratosphere by releasing enough sulphates or salt crystals to reflect a lot more of the energy from the sun back into space.

Paradoxically, there's a strong likelihood that this would actually work – based on what we've observed from the impact of massive volcanic explosions in the past reducing the amount of solar radiation reaching the Earth. It's estimated that the eruption of Mount Pinatubo in 1991, for instance, cooled the Earth by around 0.5°C for more than a year. It would also be relatively cheap. Every long-haul aircraft could be fitted with tanks and nozzles to spray the stuff around at different altitudes, and the cooling effect would be pretty much instantaneous.

But here's the problem: such a shield would *not* stop the concentration of greenhouse gases building up in the atmosphere below. So, if we ever took the shield away, by discontinuing a programme of this kind, the renewed warming effect would be equally instantaneous, and probably terminal for humankind. Environmental scientist Raymond Pierrehumbert summed up the idea of SRM as follows: 'The idea of "fixing" the climate by hacking the Earth's reflection of sunlight is wildly, utterly, howlingly barking mad.'

Worse yet, such a shield would also do nothing for the other major problem associated with the emission of greenhouse gases – the slow but inexorable acidification of our oceans. The oceans absorb around a third of the CO_2 we pump into the atmosphere, with a growing impact on corals and on many marine organisms that find it harder and harder to form and protect their shells in more acidic water. This in turn impacts food chains throughout the marine environment. A reflective shield would make absolutely no difference to this dire problem – a problem that can only be solved by dramatically curtailing the emission of those greenhouse gases in the first place.

I have to admit that I'm baring my prejudices here. I started this chapter by acknowledging that the future is entirely in our hands. For some people, that means doubling down on the kind of technology-driven fantasies that got us into this fix in the first place – fantasies that can be threaded all the way back to the time of the Enlightenment when the essence of today's still dominant model of progress was being shaped by the likes of Francis Bacon, intent on 'imposing the Empire of Man over creation'. Since then, Western civilisation has been dominated by would-be geo-engineers of every kind, attempting to control the forces of nature and to bend Gaia's will to their own.

Today's world leaders wouldn't necessarily use such language, but they would broadly subscribe to the same notion of 'human supremacy over the natural world', convinced that 'solutions' to today's Climate Emergency can only come from yet more ambitious, expensive, technology-driven innovation. Bill Gates's new book, *How to Avoid a Climate Disaster*, epitomises that kind of 'command and control' mindset, with not the slightest acknowledgement on his part that we would be well advised to learn some lessons from the last 250 years of dominating and manipulating the natural world entirely to suit our own short-term interests.

Disturbingly, those who advocate for the hard 'doubling down' geoengineering technologies are the very same people who've spent decades denying that climate change was happening – the same right-wing think-tanks; the same senior executives in oil and gas companies; the same billionaires who've worked so hard and so successfully to protect their wealth and power. I find it astonishing, therefore, that so many people can be so easily persuaded by technology-driven, get-out-of-jail-free cards of this kind.

In his excellent book on geoengineering, *Earthmasters: The Dawn of the Age of Climate Engineering*, Clive Hamilton lays bare just how devious and untrustworthy many of these geo-engineers are, not just in cosying up to philanthropists like Bill Gates and countless politicians, but in trying to introduce proposals for regulating the world of geoengineering so that they themselves end up in control – the foxes for ever in charge of the chicken coop. Hamilton argues that what they really dread is a system of transparent, rigorous and democratically accountable regulation: they want to be able to stitch things up between them with vague 'voluntary agreements', secret deals to trial a bit here or experiment over there, or 'friendly' national regulatory bodies to push plans forward without any international consensus. Any idea of a properly consti-tuted international regulatory body, under the Framework Convention on Climate Change and the ultimate aegis of the UN, is anathema to them.

I know that to some people this will sound like I'm indulg-ing myself in a really juicy conspiracy theory, and that things are now so grave that we have no time for such scruples and chronic caution if we're going to prevent runaway climate change. But if we don't get these regulatory issues properly sorted now, the consequences (both intended and unintended) could be seriously problematic further down the line.

GEOENGINEERING DONE WELL

So let's just put aside all those Solar Radiation Management fantasies, while acknowledging that we absolutely cannot turn our backs on the need for a lot more research into greenhouse gas removal technologies, as explored in the first half of the chapter. That research should be conducted with one non-negotiable condition always uppermost in our minds: that this must have no impact on the overarching priority of preventing the emission of greenhouse gases in the first place. We may well need to find ways of 'turning down the thermostat' on top of that, especially if we don't see concentrations of CO_2 in the atmosphere starting to decline from 2025 onwards. But these measures will need to be additional to, not instead of, our accelerated decarbonisation endeavours.

When it comes to removing greenhouse gases from the atmosphere, I have a loose triage system in my mind:

1. Start doing it now – *AT SCALE!*

- Accelerated tree-planting campaigns
- Forest restoration programmes
- Wetland and peatland restoration
- Restoring mangroves, salt marshes, kelp and seagrass beds
- Alternative grazing systems ('rotational grazing')
- Soil management regimes ('regenerative farming')
- Agroforestry programmes
- Fast-growing algae in open ponds

There are complex ecological challenges associated with all of these, but each has the potential simultaneously to help restore degraded ecosystems and protect biodiversity.

2. Trial it now – *AT SCALE!*

- Ocean fertilisation
- Open-ocean seaweed cultivation
- Direct air capture
- Carbon capture and storage
- Biochar (a charcoal-like substance derived from burning biomass in an oxygen-free combustion process)

3. Continue research into:

- Bioenergy with CCS (BECCS)
- Spreading finely ground silicates on both land and ocean ('enhanced weathering')
- Liming reefs and sensitive coastlands (to counter acidification)

As I said at the start of this chapter, there's geoengineering and there's geoengineering, and we need to get super-smart about the differences just as soon as possible. Even as we do everything we can to drive radical decarbonisation programmes, we have to accept, whether we like it or not, that we will also have to get billions of tonnes back out of the atmosphere to have any chance at all of staying below that 2°C threshold. And the best way of doing that will be to deploy all those natural climate solutions as fast as possible – in effect, *recarbonising* our soils, wetlands, marine environments and the rest.

Decarbonisation *and* recarbonisation; technology *and* biology; technosphere *and* biosphere – all working together in a mutually reinforcing global rescue programme.

13

PEAK MEAT

*'Activism is not a journey to the corner store, it is
a plunge into the unknown. The future is always
dark. But that darkness can be the darkness of the
womb or the darkness of the grave.'*

REBECCA SOLNIT

Sorting out the difference between 'geoengineering done
well' and 'geoengineering done heedlessly' is all about getting
back to nature: rediscovering our total dependency on the natural world; working *with* nature to create wealth, rather than
seeking progress by making war on nature; bringing nature
back into our predominantly urban and increasingly digital
lives; ensuring that all our children can grow up with at least
a bit of nature in the mix, beyond those precious early years
when every child is instinctively 'linked in' to nature before
the technosphere takes over.

I have some 'skin in the game' here, as they say. At the
start of my career, I taught for ten years in a state school in
the west of London, with some grim and soulless housing

estates as our principal catchment area. It wasn't much, but a group of us used to organise visits for some of the children to a youth hostel in Wales, enabling them to experience for a few days a bit of countryside, silence and starlit nights. Though I hate turning everything into some kind of psychological syndrome, what's been described as 'nature deficit disorder' really does hit the nail on the head here. During those visits to Wales we would also spend some time with a couple of local farmers – an eye-opening experience for most of those young people who had grown up and spent most of their lives in the city. I've remained convinced ever since that the food we eat (how it's produced, where it comes from, its impact on the environment, etc.) remains by far the most effective way of reconnecting both young and old to the natural world.

From the point of view of the Climate Emergency, food production and farming are a massive part of the problem. And dealing with that problem could be a massive part of the solution. For all sorts of reasons, it took the IPCC a very long time to get round to looking in detail at the impact of agriculture on the climate – with its first major report in this area in October 2018. The headline was striking: 'Land use contributes about one-quarter of global greenhouse gas emissions, notably CO_2 emissions from deforestation, methane emissions from rice and ruminant livestock, and N_2O emissions from fertiliser use.' Depending on how you do the breakdown, livestock accounts for between 15 and 18 per cent of global greenhouse gas emissions. So this is *huge*.

It's also about land use as much as it is about greenhouse gas emissions. Roughly 50 per cent of the Earth's 'habitable land' is used for agriculture; 77 per cent of that is used for livestock, including the production of animal feeds, with 23 per cent devoted to crops for direct human consumption. And here's the eye-catcher: protein is the most important part of balanced, nutritious diets, but just one-third of the protein we consume comes from that 77 per cent of habitable land used for livestock, with two-thirds coming from the remaining 23 per cent used to produce food for direct human consumption. That's how crazy things have got.

The livestock industry must also take responsibility for a host of additional environmental issues as well as greenhouse gas emissions, in terms of its massive impacts on soil, biodiversity and water. Agriculture is responsible for around 70 per cent of all freshwater use in the world today, and the livestock industry is responsible for about a third of that. In a world where the lives of more and more people are already profoundly affected by water scarcity, this is going to become more and more of a problem. According to the UN, 1.8 billion people will be living in countries or regions with absolute water scarcity by 2030 – with up to two-thirds of the world's entire population living in countries suffering from water stress. How long before people start joining the dots between water used so wastefully for livestock production, and chronic, life-threatening water shortages?

It gets worse. A new report from the think tank Chatham House, endorsed by the UN, shows that conversion of natural ecosystems for crops or grazing has been the biggest single driver of habitat loss over the past 50 years. And it's now the biggest single risk to the 28,000 plants and animal species known to be at risk of extinction – all driven by the overproduction of crops and cattle to ensure that meat-based products remain as cheap as possible. Lead author Professor Tim Benton put it as follows:

> Politicians are still saying 'my job is to make food cheaper for you', no matter how toxic it is from a planetary or human health perspective. We must stop arguing that we have to subsidise the food system in the name of the poor, and instead deal with the poor by bringing them out of poverty.

It's not all about raw numbers. There are deep ethical issues here in terms of animal welfare and our apparent readiness to go on inflicting unspeakable cruelty on billions of creatures (70 billion in 2018) reared in factory farms of one kind or another, where the animals never see the light of day. Around two out of every three farm animals are factory-farmed. In the USA, this figure is 99 per cent. While some of these factories operate to notionally 'humane' standards, the majority (on a global basis) absolutely don't. There's now a huge amount of written and filmed material out there highlighting the horror stories behind factory farming.

By any measure, the livestock industry is a massive net contributor to today's Climate Emergency, and the faster we can see global meat consumption peak, and then start coming down, at the same sort of speed as emissions from burning fossil fuels, the safer we will all be. As you might imagine, that was a big area of inquiry for me in writing *The World We Made* back in 2012, looking back from 2050 at what it was that enabled us to avoid runaway climate change. This was my speculation at the time:

Back then, the experts were confidently predicting that meat consumption would grow from around 290 million tonnes in 2010 to around 460 tonnes in 2050. Well, that didn't happen: instead, public opinion began to change, for both health and environmental reasons. Per capita meat consumption plateaued in 2030 at around 355 million tonnes in total – the "peak meat moment", if you like!

So, eight years on, just how good were my forecasting instincts on this particular sustainability indicator? The jury, I'm afraid, is still out. According to the Food and Agriculture Organization (FAO), total meat consumption in 2018 increased to 340 million tonnes, but declined significantly in 2019, and is forecast to have declined again in 2020 (possibly by as much as 3 per cent, not least because of COVID-19). Two consecutive years of decline would seem to indicate something of an emerging trend, even though it might still go up again in 2021. With nine years still to go, I'm sticking to my guns, especially as we are now beginning to witness much more profound changes in consumer behaviour.

Eat less meat – eat more plants!

There's been a marked increase in the numbers of people in the rich world who have become vegetarians or vegans over the past few years, be that for health reasons, animal welfare reasons or environmental reasons. Around 3 per cent of US citizens are now vegans, and a further 5 per cent are vegetarians, according to a 2018 Gallup poll. In Australia, currently just 5 per cent of the population identifies as vegetarian or vegan, with many more women than men now avoiding meat in these countries.

As far as the UK is concerned, a survey carried out by Finder in February 2021 revealed more and more citizens are following a meat-free diet, with 3 per cent self-declaring as vegans, 6 per cent as vegetarians and nearly 5 per cent as pescatarians (fish only). That's around 7 million people, with a further 6.5 million indicating that they too wanted to become meat-free in 2021. Younger generations are significantly more likely to adopt meat-free diets, with 20 per cent of Gen Z already doing so. On top of that, there are far more people actively reducing the amount of meat that they eat every week.

Although it's extremely difficult to get consensus on the actual figures, it's clear that the plant-based movement has now gone well beyond a temporary fad or market blip. More and more people in the West recognise how important it is that we should seek to reduce average meat consumption, taking advantage of the extraordinary diversity of high-quality, plant-based products now coming to market. However, meat-eating elsewhere in the world is still growing, in line with people's improved standard of living. As the World Health Organization says: 'There is a strong positive relationship between the level of income and the consumption of animal protein, with the consumption of meat, milk and eggs increasing at the expense of staple foods.'

In the cut and thrust of this crucially important debate, there's always a personal dimension. I'm not a vegetarian, but I don't eat much meat these days, and when I do, it's because I trust its provenance or might even be lucky enough to know the farmer selling it. As a lifelong supporter of Compassion in World Farming, it's the intensive factory farming that I feel so strongly about, both from a welfare and a sustainability point of view.

For me, therefore, mixed farming, committed to high welfare, extensively grazed livestock integrated into a largely organic or regenerative farming system, will always have a case to make. Interestingly, there are more farmers now who are interested in bringing some livestock back into their complex arable rotations as a way of providing more 'home-grown' nutrients, thus reducing the need for synthetic fertilisers. And there's now much more research being done on 'silvopasture', the benefits of which are captured in Project Drawdown's summary:

> From the Latin for 'forest' and 'grazing', silvopasture is just that: the integration of trees and pasture or forage into a single system for raising livestock, from cattle and sheep to deer and ducks. Silvopastoral systems sequester carbon in both the biomass above ground and the soil below. Pastures that are crisscrossed with trees sequester 5 to 10 times as much carbon as those of the same size that are treeless.

I really don't think that today's crude polarisation ('animal foods bad, plant foods good') is particularly helpful. We've learned the hard way that setting out to 'shame' people into changing their behaviour, whether that be in terms of meat consumption, flying, or 'retail therapy', can have very mixed results, with as many people reacting negatively – and often doubling down on that behaviour – as reacting positively.

Policy dilemmas

The truth of it is that governments are very unlikely, anytime soon, to be setting out to meet their commitments under the Paris Agreement as much through reductions in meat consumption as through reductions in burning fossil fuels. It's become one of those ludicrous no-go territories in a world of deregulatory, hands-off governments, where fear of gratuitous 'nanny state' accusations from right-wing media outlets counts for a lot more than smart, forward-looking legislation. Instead, everything remains voluntary and advisory, with worthy guidelines to help inform and educate consumers.

Worse yet, governments are as much 'in hock' to today's intensive, life-destroying agribusinesses as they are to today's life-destroying fossil fuel businesses, particularly in the US. A report in September 2019 from the Food and Land Use Coalition in the UK estimated that governments are spending between $700 billion and $1 trillion of taxpayers' money a year on subsidies for agriculture, which works out at $1 million a minute. A tiny fraction of that (around 1 per cent) is used to promote more environment-friendly farming, but the rest goes on carbon-intensive commodity crops, livestock farming, clearing forest, and fertiliser and water subsidies, all of which further exacerbate today's Climate Emergency and ecological crisis. And it's primarily big farmers who benefit the most.

The report specifically rejected the idea that subsidies are still needed to ensure supplies of cheap food for consumers on low incomes, with the cost of the damage done by modern agriculture estimated to be greater than the value of the food produced. Its principal recommendation was to redirect these subsidies into storing carbon in soil, cutting food waste, planting trees and so on (as I will cover in Chapter 17), which would also provide continuing support for farmers doing the right thing.

This is simply the latest in a whole series of blockbuster scientific reports demonstrating just how broken the world's food and farming system really is, with billions of people either undernourished or overweight/obese. Back in 2015, the FAO set out to highlight just how dysfunctional things are: 'In many countries, there is a worrying disconnect between the retail price of food and the true cost of its production. As a consequence, food produced at great environmental cost in the form of greenhouse gas emissions, water pollution, air pollution and habitat destruction, can appear to be cheaper than more sustainably produced alternatives.'

Which brings us straight back to that old story of 'cost internalisation': how best to reflect the true costs of production in the price we pay for things, even if that means challenging conventional wisdom about the benefits of keeping things as cheap as possible for consumers. When it comes to modern farming, there is a long list of costs that need to be taken into account: damage to soil through poor farming practices; overuse and pollution of water; loss of biodiversity; emission of greenhouse gases; the overuse of nitrogen-based fertilisers; impacts on human health, including food-borne diseases, obesity and antimicrobial resistance. We're talking billions and billions of dollars, euros or pounds, with none of it actually reflected in the price we pay for our food, but paid for nonetheless by society or by future generations.

In 2017, the Sustainable Food Trust carried out a comprehensive analysis of what this means in practice for UK consumers. 'The Hidden Cost of UK Food' came to the following conclusion:

> UK consumers spend £120 billion on food every year, yet there are serious environmental and health-related costs that generate a further £116 billion of costs. Significantly, these costs are not paid for by food businesses and the farming practices that cause them, nor are they included in the retail price. Instead, they are being passed on to the public through taxation, lost income due to ill health, and the cost of mitigating and adapting to climate change and environmental degradation. In effect, this means that UK customers are paying almost twice for their food, despite being told by the media that food has never been cheaper.

(Since 2017, the Sustainable Food Trust has revisited some of its calculations, with the cost of some elements being revised downwards, and others being revised upwards, with a view to bringing out a new version of the report in 2021.)

But governments are understandably nervous about measures that impact disproportionately on poorer people by raising the price of food – recent history is awash with examples of governments swept away by food riots or intense political protest. Blanket taxes on livestock production would almost certainly be regressive in that regard, even if the revenue raised from such taxes was to be used either to subsidise healthier food, or to encourage farmers to produce to higher environmental and animal welfare standards. Other fiscal measures could be much more carefully targeted specifically to address certain problems – excessive use of nitrogen-based fertiliser, for instance, or the use of antibiotics in intensive livestock systems.

Antimicrobial resistance is already a massive problem, and although this is caused more by overprescribing for human health reasons, the continuing abuse of antibiotics in factory-farming systems is a very big part of the problem. As far back as 2015, the US Centers for Disease Control and Prevention came to the astonishing conclusion that at least 2 million people in the US are infected with antibiotic-resistant bacteria every year, causing at least 23,000 deaths. The direct cost of this is as high as $20 billion a year.

In February 2019, Professor Sally Davies stepped down from her role as Chief Medical Officer (CMO) for England, and is now the UK's Special Envoy on Antimicrobial Resistance. Nobody's been more outspoken about the 'double whammy' going on here: very few new antibiotics have been developed over the past twenty years (caused by what she described as a 'discovery void'), at the same time as many diseases are evolving and becoming more resistant to existing drugs.

During her time as CMO she commissioned a study in 2014 showing that a continued rise in resistance by 2050 could lead to the deaths of at least 10 million more people worldwide every year, having warned earlier what that might look like:

Antimicrobial resistance poses a catastrophic threat. If we don't act now, any one of us could go into hospital in twenty years for minor surgery, and die because of an ordinary infection that can't be treated by antibiotics. Routine operations like hip replacements or organ transplants could be deadly because of the risk of infection.

Her mission today is absolutely critical: to persuade people that antibiotic resistance poses a threat to future generations as big as climate change. However, huge numbers of people remain indifferent to the scale of this threat, often demanding that they should be prescribed antibiotics even when they know they will be completely ineffective.

'CLEAN MEAT'

The use of antibiotics in factory farming is already a big issue (contributing directly to antimicrobial resistance), and is therefore likely to become a decisive factor in persuading hundreds of millions of people to start choosing 'artificial meat' products rather than 'real meat' products. This is already a big industry, sometimes referred to as cell-based meat, lab-grown meat ('labriculture'), clean meat and cultured meat (in comparison to plant-based meat alternatives). The basic technology (taking tissue from an animal, using special cultures to allow the cells to multiply in a fermenter or a bioreactor, and gradually building up muscle tissue) is being refined all the time, ensuring the costs (which were astronomical in the early days) are also starting to come down. As Patrick Brown, CEO of Impossible Foods, puts it: 'Unlike the cow, we get better and better at making meat every single day.' One of the most interesting indicators of potential breakthroughs here is the astonishing amount of money going into today's clean-meat trailblazers (Memphis Meats, Beyond Meat, Impossible Foods, JUST, Mosa Meat, SuperMeat, Mission Barns, New Age Meats, etc., etc.) as well as into their 'clean-fish' equivalents (Finless Foods, Wild Type, BlueNalu) – from venture capitalists, philanthropists like Bill Gates, and even from large meat companies themselves.

Right now, only a relatively small number of people have ever tasted a clean-meat product, but this is going to change dramatically over the next few years. In September 2019, a new report from the UK-based think tank RethinkX, 'Rethinking Food and Agriculture 2020–2030', set out to demonstrate just how quickly a convergence of new technologies could fundamentally disrupt every single aspect of today's meat, fish and dairy industries. Key to this is the revolutionising of the very old technology of fermentation through the use of 'precision biology', bringing together machine learning and artificial intelligence with different kinds of biotechnology to produce almost any complex organic molecule.

Let's just demystify all this turbocharged terminology. Consider how the cow turns all that grass it eats into protein (casein and whey) thanks to the intervention of trillions of productive microbes in its rumen. Then imagine that you didn't need to do that in the cow's rumen itself, but could use those productive microbes to produce the same functional proteins (casein and whey) in a fermenter instead of a cow. Those proteins make up just 3.3 per cent of milk's overall composition – the rest is water (87.7 per cent), lactose (4.9 per cent), fats (3.4 per cent), and some minerals and vitamins. It's only the 3.3 per cent that precision fermentation has to substitute for – the rest can be added later.

Is this for real? Ask any diabetic. Historically, diabetes was treated with insulin extracted from the pancreases of cows and pigs, and then processed (very expensively) to reach the level of purity required for human use. In the late 1970s, a company called Genentech produced a genetically modified yeast capable of producing human insulin – revolutionising diabetes treatment through one of the first examples of precision fermentation. The authors of 'Rethinking Food and Agriculture' are in no doubt about the scale of disruption coming down the track:

> We are on the cusp of the deepest, fastest, most consequential disruption in food and agricultural production seen since the first domestication of plants and animals 10,000 years ago. The cost of proteins will be five times cheaper by 2030 and ten times cheaper by 2035 than existing animal proteins. This means that, by 2030, modern food products will be higher quality and cost less than half as much to produce as the animal-derived products they replace. Industrial food production systems have as much chance of competing with modern foods as cuneiform clay tablets had of competing with modern computer tablets or smartphones.

Forget all that stuff about people being unable to overcome the 'yuk factor' associated with cell-based meat. In future, that may remain the case with a small number of people squeamish about the idea of 'meat grown in labs', or with 'real men demanding real meat', but let's be logical here: clean meat is still an animal-based product. And it's not only going to taste just as good, it's going to be cheaper (as economies of scale kick in), healthier (a clean burger can contain not only less fat and salt than a cow-based burger, but more vitamins and minerals than a portion of fresh vegetables), safer for *all* of us whether we eat meat or not (no antibiotics are needed in a modern fermenter, which will significantly reduce total global usage of antibiotics); with far lower emissions of greenhouse gases and zero impact on water, soil and biodiversity. Apply this across the entire intensive livestock industry, and roughly 70 billion animals will not need to be raised in factory farms and slaughtered every year to satisfy our flesh-eating habits.

I see this clean-meat revolution as the exact equivalent of the clean-energy revolution: massive beneficial impacts, delivered at scale and at speed, in ways that ensure a just and fair transition for rich and poor alike, with people needing to pay less for more balanced, nutritious diets. In the UK, for instance, campaigner George Monbiot has calculated that 'If our grazing land was allowed to revert to natural ecosystems, and the land currently being used to grow feed for livestock (55% of the UK's cropland) was used for grains, beans, fruit, nuts and vegetables for humans, this switch would be equivalent, all together, to absorbing nine years of our total current emissions.'

Personally, I don't believe we need such an absolutist solution for the UK, where mixed use, organic and regenerative livestock farming will still have an important part to play. But this does mean that 'real' meat will cost more (but we'll all be eating a lot less of it, and many more people won't be eating any of it), while 'clean' meat will cost a bit less.

And then imagine the global consequences of similar transformations in those parts of the world where per capita meat consumption continues to rise every year, particularly in Asia. This means that demand for animal feeds is also continuing to grow – according to some estimates by around 4 per cent per annum from 2020 through to 2030. By any measure, this cannot possibly be reconciled with addressing today's Climate Emergency.

The Chinese are big meat-eaters: China consumes around 28 per cent of the world's meat, with roughly 18 per cent of the world's population. There are some early indications that consumer trends are now shifting, with more vocal advocacy from both vegetarian and vegan organisations, with the market for plant-based meat alternatives growing fast. However, coming from a very low base, sales have still not reached $1 billion a year, and all these plant-based alternatives cost much more than real meat products.

So here's an offbeat thought to end this chapter: what if China was to become the most ambitious driver of the global clean-meat revolution, surpassing countries like the USA and Israel (current market leaders in this space), just as it has already done in the clean energy area? China currently imports around 80 per cent of Brazil's soybeans (at around $23 billion a year) for animal feed, and is increasingly desperate to find other overseas sources so that it can protect its own limited land to produce food for *direct* human consumption. China hates being dependent on foreign producers; China knows that its citizens are not going to be told to stop eating meat; and China knows that today's Climate Emergency is going to be devastating for hundreds of millions of those citizens unless rapid action is taken on a global basis.

There are not many ways of squaring that particular circle – which is why I would put any amount of money on China being *the* leader in precision fermentation and cell-based technologies by the end of the decade.

14

CHINA: HEADING FOR THE ABYSS

'Rebellions are built on hope.'
STAR WARS

Between 1958 and 1961, somewhere between 20 and 35 million Chinese died in the so-called 'Great Famine'. Chairman Mao's Great Leap Forward had forced many peasants to leave farming behind and take up jobs in the emerging iron and steel industries; a whole battery of ill-conceived policies intended to reshape agriculture compounded the resulting crisis. Chinese leaders in the modern age were all scarred by that trauma, ensuring that the widely used notion of 'food security' has been an imperative for policymakers in a way that few other countries can imagine. Feeding 20 per cent of the world's population with just 7 per cent of the world's land remains a massive challenge.

Which means that every year's increase in per capita meat consumption in China is seen as another small triumph. In the 1960s, the average was around 5kg of meat per person per annum, rising to around 20kg in the 1980s. These days it's around 60kg and still rising, compared to around 115kg in the

US and 80kg in the UK, with consumption now falling in both those countries. So China's consumption is still around half what the average American consumes every year. Many commentators now doubt that this gap can ever be eliminated – primarily because China will not be able to import enough soybean to feed the growing number of livestock. China once grew almost all the soybean it needed itself, but from the mid-1990s onwards that changed. Now it produces only 16 million tonnes of the 110 million tonnes it needs, with the rest imported from the US, Brazil (putting at risk huge amounts of land in both the Amazon and the Cerrado for further soybean production), and Argentina.

It's easy to be critical, but the Chinese argue that they have as much right to increased meat consumption as any other country. It's still the most compelling proxy for a higher material standard of living, symbolising China's extraordinarily rapid escape from crushing poverty – it's as completely logical an aspiration as it is completely unsustainable. No one is more aware of this than Chinese policymakers themselves – which may explain why their dietary guidelines in 2016 recommended per capita consumption of no more than 27kg per annum, roughly half where it stands today.

Trade-offs of this kind are commonplace in a nation of around 1.4 billion people, a population still growing and still hungry for all the trappings of twenty-first-century consumerism.

CHINA'S GROWTH MACHINE

Since 1979, China's GDP has grown by an average of just under 10 per cent per annum. Even now, President Xi Jinping will not allow economic growth to fall below 6.5 per cent, making this police state (run by 90 million members of the Communist Party) one of the biggest drivers of ecological collapse and runaway climate change in the world today.

As China became 'the workshop of the world' from the mid-1980s onwards, Western consumers were the primary beneficiaries as prices fell on products of every kind. The West lost millions of jobs in the process (leading directly, as many have argued since, to stagnating living standards, intense disillusionment on the part of tens of millions of 'left behind citizens', and the resulting election of Donald Trump and the UK's disastrous exit from the EU). But the principal casualty of this 'hyper-industrialisation' was the environment in China, with horrendous impacts on rivers and lakes, on aquifers, on soils and farmland, on air quality, on its forest and wilderness areas. As described in Richard Smith's devastating new book, *China's Engine of Environmental Collapse*, it is almost impossible to overstate the cumulative impact of this – not just on the environment itself but on people's health and life expectancy.

There are ways to stop toxic pollution. But these cannot be implemented in an economy where maximising growth is the highest priority. Many of China's filthy industries are impossible to clean up in any meaningful sense, and many, if not most, are dedicated to producing products that neither the Chinese nor anyone else really needs, that should never have been produced in the first place.

China is also by far the greatest threat to global biodiversity. It is the principal driver of deforestation in South-east Asia and elsewhere, and the largest importer of illegally-logged timber and illegally-poached and trapped wildlife. Its obsession with totally ineffective 'traditional medicines' (praised to the skies by Xi Jinping himself as 'a gem of ancient Chinese science') means that China rides roughshod over all international regulations, sucking in resources from all over the world as some kind of giant vacuum cleaner.

Little of this is seen by people either outside or even inside China – the Government's near-total control of the media (including social media) censors anything seen to be critical of the Communist Party and of Xi Jinping himself. The exception to that rule is the appalling air pollution problems that hundreds of millions of Chinese face every day, caused both by emissions from coal-fired power stations and by vehicle exhausts.

In 2015, Chinese journalist Chai Jing, fearful for the health of her six-year-old daughter and for all Chinese children, produced a powerful documentary called *Under the Dome*, highlighting not just the astonishing scale of the problem (with thousands of people dying every day as a consequence of chronic air pollution), but the complete failure of policymakers to get on top of the problem. It was a runaway success – until, almost inevitably, it was suppressed by the authorities.

But this genie is never going back in its bottle. Public concern remains high, even as things have started to improve. Access to information has massively benefited from this high profile, with a number of apps providing accurate and real-time data on key pollutants – particularly the $PM_{2.5}$ particles that are such a threat to human health. Government officials are under no illusions about the significance of this air quality crisis, seeing it as one of the greatest potential threats to social stability in the future.

China's 'growth-at-all-costs' model of development is not just condemning its own citizens to an unprecedented environmental reckoning in the near future, but condemning the whole of humankind to a climate meltdown of an almost inconceivable magnitude. While the rest of the world starts to confront its responsibilities in the midst of the Climate Emergency, China plans to continue on exactly the same suicidal course it's been on for the last 40 years. And that's all about coal.

It's China's access to almost limitless amounts of cheap coal that has driven its so-called 'economic miracle' over the past forty years, and which still (supplying around 65 per cent of electricity demand) underpins its economic prospects. China became the largest emitter of greenhouse gases back in 2006, and today's emissions amount to a whopping 30 per cent of the global total. The per capita figures tell an interesting story: the average citizen in the USA and Australia emits around 16 tonnes per annum. In China, it's 7, but still rising. In the UK, it's 6, and still falling.

There's much talk of China's ambition to reduce dependence on coal – as there has been for years. Most of it is complete nonsense. Its existing commitments are as minimalist as they can get away with: emissions peaking in 2030 (i.e. going up and up until then); modest reductions in the amount of CO_2 emitted per tonne of output; and a 'Net Zero economy' by 2060. It is of course possible that China will come forward with more ambitious plans before COP26 at the end of 2021, but its latest Five-Year Plan (released in March 2021, covering the 2021–2025 period) provided little reassurance on that score. Emissions continue to rise year on year – even in 2020, the year of the pandemic, they increased by nearly 1 per cent, according to the International Energy Agency.

The reality is that the over-arching goal of maximising economic growth in the short term, in its obsessive desire to catch up with and then overtake the USA as the world's largest economy, matters far more to President Xi Jinping and the Communist Party than avoiding runaway climate change in the long term.

And that's somewhat ironic. There's no debate about the science of climate change in China; many of its politicians are scientists and engineers, and are in no doubt that the impact of accelerating climate change on China's economy will be dire. They're especially fearful of the impact of drought and desertification (with terrible dust storms in Beijing in March 2021), of extreme heat on its farmers, and of rising sea levels putting at risk many of the biggest and most important cities on its eastern seaboard. Some of the most nightmarish projections of the consequences of a 1 metre sea level rise by the end of the century involve the sprawling megacity of Shanghai, which is already experiencing significant problems as a result of saltwater incursion into its aquifers.

What makes this all the more reprehensible is that so much of China's economic growth serves no purpose other than to keep the Communist Party in power by providing millions of people with jobs in construction. *China's Engine of Environmental Collapse* provides an astonishing picture of 'overproduction' at every level: ghost cities (sixty-five million urban apartments remain unoccupied), all-but empty high-speed trains and stations, four-lane highways that go nowhere, including thousands of unnecessary bridges and tunnels, under-utilised factories and chemical plants, and so on. Much of this is driven by competition for status between Party bosses at the provincial and municipal level – on whom Xi Jinping's ruthless, sociopathic writ would appear to have little impact.

China's 'green revolution'

All of which provides the backdrop to what can only be described as China's very own renewables revolution. It's hard to exaggerate the impact of this on the world's energy prospects – and, directly through that, to be fair, on the world's prospects of avoiding runaway climate change. China doubled its already huge investment in renewables in 2020. It's installed far more wind power and far more solar power than any other nation on Earth. Five out of the ten largest wind power companies in the world are Chinese; nine out of ten of the world's largest solar companies are Chinese – and it's this concentration of manufacturing and R&D muscle that has so dramatically driven down the cost of both wind and solar year after year after year.

I looked at the global implications of that in Chapter 4, but the solar story in China itself is truly remarkable. Back in 2008, domestic Chinese solar was a bit-part player, with almost all production exported to Europe and the US. After the financial crash, those export markets slowed right down, leaving the industry with huge unused manufacturing capacity. The Chinese government decided then and there to build up its domestic industry; in 2007, China had just 100MW of installed capacity; by 2017, that had grown 130-fold! Most of that comes from huge ground-mounted solar farms, covering an area of around 1,500km^2, approximately the size of Greater London.

However, we have to keep all this in perspective. Just go back to the numbers on page 45. Most of China's renewable electricity still comes from hydropower, with less than 6 per cent coming from wind and solar. The good news is that the investments will continue. In just three years, the International Energy Agency estimates that China's installed solar capacity will exceed 320GW – which, just to get it in perspective, happens to be the total electricity demand of Japan. We're not talking small beer here.

But nor are we talking about flawless execution. The capacity may have been installed, but it isn't necessarily connected to the grid: 70 per cent of the new capacity is out in the western provinces of China, with most of the demand over in the east, as is the case with many of their biggest wind farms. They're only now installing the kind of ultra-high-voltage transmission lines needed to connect the two – to the tune of around $100 billion of new investment.

What's more, regional governors (often with powerful coal interests of their own) aren't as enthusiastic about renewables as Beijing, frequently dispatching electricity from coal-fired power plants onto the grid before any green electrons. In 2019, the Chinese government cracked down on this by introducing a 'green-first' dispatch mandate – though it's already clear that this is being routinely ignored by local Party bosses. China watchers are fascinated by this 'variable geometry' at the regional level, with some regions going hell for leather to be in the vanguard of the green revolution, and others hanging on like grim death to coal mining, iron ore, steel production and the rest.

Although we in the West see China as one huge top-down autocracy, it's not like that on the ground. What's more, there's deep and persistent corruption at every level – despite Xi Jinping's efforts to crack down on this, which he sees as a highly problematic constraint on his ability to rule absolutely.

However fast renewables may be growing, from a still very low base, it will make little difference over the next decade if emissions from coal-fired electricity continue to rise. And that's a critical factor in China's transition away from the internal combustion engine towards electric vehicles – if the electricity that those EVs are using is provided predominantly by dirty, highly inefficient coal-fired power stations, then it's really not much of a win for our low-carbon prospects.

But it will happen. If increased meat consumption is an important measure of rising prosperity in China, then getting a car represents 'peak aspiration'. At the end of 2019, there were already 280 million vehicles on China's roads, although that is still a low per capita figure: in the US, there are 840 registered vehicles per 1,000 citizens, whereas in China it's just 200. But that's where this transition story gets so interesting. China became the largest market for electric vehicles in the world in 2015; in 2017, it accounted for half the total global sales of EVs; in 2018, more EVs were sold in China than in the rest of the world combined!

There are now 4.5 million EVs on China's roads. It has an aspirational target of 100 million by 2030. There are already

100 electric car manufacturers in China, with thousands of companies involved in this burgeoning supply chain. So they're not hanging around – in comparison to all of us slowcoaches in the West. This is one area where one can talk uncontroversially about China seeking global domination – and not just with EVs. There are already 250 million electric bikes on China's roads, and around 400,000 electric buses. 100 per cent of Shenzen's 16,000 buses are already fully electric. China produces 99 per cent of all electric buses in the world.

Not surprisingly, Western companies are struggling to keep up. Having already spent more than $60 billion driving this revolution, including generous subsidies for EVs, with another $60 billion to be spent over the next ten years, the Chinese government is now developing a timetable to ban the manufacture and sale of any petrol or diesel cars at some point in the future – and there is a growing number of pundits who believe they will be opting for 2030. That is a bona fide game-changer – partly driven, it has to be said, by China's strategic priority of reducing its dependency on imported oil.

As we saw in Chapter 4, the story of EVs is really a story about *batteries*. Because he's such an astonishing self-publicist, you probably think that Elon Musk, of Tesla fame, is the go-to guy when it comes to batteries, with all his talk of his 'Gigafactory' in Nevada. Forget it. The Chinese are already responsible for two-thirds of global lithium-ion battery manufacture; the USA just 10 per cent. In terms of the raw materials required for battery manufacture, the Chinese now control between 50 and 70 per cent of global supply. And with all this industrial muscle deployed so aggressively, they've been almost as effective in bringing down the costs of battery production (down 60 per cent over the past five years) as they have been with solar panels. Tesla is itself about to open a huge new factory in China, significantly ramping up competition with Chinese manufacturers.

There are huge problems associated with this transition. This is China: there is little consideration given to environmental impacts, particularly when it comes to demand for cobalt, a critical metal in battery manufacture. China Molybdenum now controls more than 15 per cent of cobalt production, with major investments in the Democratic Republic of Congo, already a notorious hotspot for flagrant abuses of human rights. Chinese companies of one kind or another now control more than 85 per cent of the world's refined cobalt capacity.

At every point in this story it is important to realise that China never loses sight of its strategic intent to use renewables to help dominate critical sectors of the global economy. And nowhere is that clearer than with the extraction and refining of so-called 'rare-earths', concentrations of precious metals that have remarkable properties of huge value to the world's renewable energy, batteries and IT sectors. Demand for these rare earths is growing exponentially, and China currently controls at least 90 per cent of production.

And then there's lithium. China (or, rather, Tibet) may not have the largest reserves of lithium – that's Chile – and it's not even the largest producer of lithium – that's currently Australia. But it soon will be. Use of lithium-ion for batteries in China is projected to increase tenfold by 2030, and Chinese companies are expanding the mining and extraction of lithium from Tibet at an astonishing rate, causing horrendous environmental pollution in some of the world's most sensitive (and most sacred) places.

THE ABUSE OF POWER

The extraction of lithium represents but a small part of China's continuing exploitation of Tibet's natural resources – and a small part of its subjugation of the rights and the culture of Tibetan people. This seventy-year horror story is now compounded by its ruthless oppression and ethnic cleansing of the Muslim Uighurs in neighbouring Xinjiang, anywhere between a million and three million of whom are now being held in so-called 're-education camps'. These look pretty much like the Nazi concentration camps in the Second World War, minus the gas chambers.

In the opinion of many human rights activists, Xi Jinping's personal involvement in these atrocities, in both Xinjiang and Tibet, ensure that he should be prosecuted for violation of the UN's Convention against Genocide.

All this poses a serious dilemma for Western companies, given that almost every solar panel sold in the EU includes some polysilicon from the Xinjiang region – 45 per cent of the global supply of high-grade polysilicon comes from Xinjiang. It's well known that the 're-education camps' double up as forced labour camps, and there is now growing evidence that China has been intent on 'compulsory upskilling' of Uighurs in this high-tech sector.

China's fearsome determination to control the life of every single one of its 1.4 billion citizens is unlike anything else going on in the world today. It's currently introducing the most extensive network of facial recognition surveillance cameras in the world, and is pressing ahead, despite significant public concern, with a nationwide 'social credit system' to give every citizen a rating based on behaviour, political attitudes, support for the community and for government initiatives. Its formidable system of internet censorship is being exported to other autocratic regimes.

It's hard – really, really hard – to reconcile all of this with Xi Jinping's commitment to securing 'an ecological civilisation' in tomorrow's China. This concept was first aired back in 2007, and then formally enshrined in the Chinese Communist Party's constitution in 2013. But when it comes to 'balancing green mountains and gold mountains', in Xi Jinping's own words, the Chinese President always goes for gold. There is no indication that this commitment to an ecological civilisation amounts to anything other than greenwashing hypocrisy at the highest level.

In both 2015 and 2017, the government significantly 'upgraded' its Environmental Protection Law, and Western NGOs operating in China (particularly the US-based Natural Resources Defense Council, which has been providing technical support to government departments for more than twenty years, and UK-based ClientEarth, which has focused on the legal system, helping to strengthen the courts and train hundreds of Chinese lawyers) now report some movement in the way China's uniquely awful environmental legacy is being addressed. In a country where 60 per cent of fresh water is polluted (much of it so seriously that it cannot even be used for industrial purposes), where huge amounts of land are severely contaminated, where food adulteration scandals are commonplace (making Chinese citizens more aware of food quality issues than most of us in the West), and where tens of millions of hectares of productive land have been lost either through poor agricultural practices or through the most comprehensive programme of urbanisation ever seen on planet Earth, tough, timely and consistent application of environmental laws is absolutely critical.

Here again, however, the Communist Party, at the regional or local level, often sees things differently. Local companies contribute significant tax revenues to them, not to the central government, which makes them very reluctant to crack down on their illegal activities. The local courts are under their direct control. They often refuse to hear cases, or even to release critical data about pollution episodes; environmental lawyers working with local NGOs are often intimidated or detained. Even when fines are imposed, they're often not collected.

Chinese NGOs are adept at working with 'systematic ambivalence' of this kind, with organisations like the Institute of Public and Environmental Affairs, China Pollution Map and Green Choice Alliance now using the law (and China's unique political dynamics) to ensure appropriate corrective action. To a certain extent, the Beijing government actually needs them to help reinforce its central writ at the regional level, but they also understand that there is always a line out there in terms of campaigning tactics that cannot be crossed without punitive sanctions being taken against them.

We in the West will always argue the corner for the superiority of democratic processes, laws, regulations and standards, though the track record of many Western countries (particularly the UK and the US these days) is pretty patchy in practice. The idea of 'environmental authoritarianism' makes us extremely nervous. China's efforts to use the apparatus of the state to restore some sort of balance between its economy and nature will become absolutely critical in the future. But the reality is, in one of the most powerful police states in history, that there is no effective Rule of Law, either constitutionally or through the diktat of its Supreme Leader.

One area where Beijing's central mandate has been effective is the Great Green Wall, a massive tree-planting campaign that has raised the amount of forest cover in the country from around 10 per cent in 1990 to something like 20 per cent in 2016, with a plan to plant 100 billion trees by the time it's finished. Without a doubt that makes it the largest 'eco-engineering' project on the planet. Some experts believe it would be better to focus more on grasses and shrubs in some of the particularly arid areas in the north of China, but the programme has already significantly reduced the number of dust storms to which many of China's cities were prone – and sequestered a hell of a lot of CO_2 in the process.

Examples of this kind are important. For the truth of it is that China is now *the* superpower that matters most to most people in the world – in Asia, South-east Asia and the Pacific. Its reach is extraordinary. I suspect most of us are only just waking up to the true significance of China's most impactful foreign policy: the Belt and Road Initiative (BRI), a colossal infrastructure and project finance programme operating in seventy countries, with a planned cumulative investment over thirty years of between \$4 trillion and \$8 trillion. Until now, most of these investments have been made with little or no concern for the environment. Between 2001 and 2016, China helped finance 240 coal-fired power stations in twenty-five BRI countries, and has funded (or built itself) thousands of kilometres of new roads with not so much as an environmental assessment along the way.

That could be changing. In 2017, Xi Jinping issued new guidance on promoting 'a Green Belt and Road', indicating that all future investments would be subject to much more stringent risk assessments and environmental conditions. In 2017, it also invested $44 billion in renewable energy schemes across South-east Asia – schemes that may well not have got off the ground at all given how utterly deplorable investment banks in that part of the world have been in recognising their environmental and climate responsibilities. China has already committed $3 billion to support climate adaptation measures around the world – as much as the USA.

In so many different ways, China does indeed 'hold the key'. Without China continuing to lead on renewables, EVs and other green initiatives, any prospect of halving global emissions by 2030 becomes a great deal less achievable. But its continuing dependence on coal to drive high levels of economic growth through to 2030 and beyond still makes China the biggest threat to any long-term prospects for a stable climate.

There is much else about China's leadership ambitions that remains deeply problematic, and much else about its abuse of human rights that remains utterly abhorrent. It leaves many questioning just how high a price will need to be paid if China's planet-wrecking and genocidal autocracy cannot be tempered.

Part 4

WHAT'S STOPPING US?

15

LETHAL INCUMBENCIES

*'Because things are the way they are, things will not
stay the way they are.'*

BERTHOLD BRECHT

China has led the world in one further critical area: poverty
reduction. At the end of the 1970s (thirty years after the
founding of the People's Republic of China), the proportion
of China's rural population living in poverty was still an astonishing 97.5 per cent – around 770 million people at that
time – and there was also extensive poverty in its towns and
cities. According to the World Bank, the number of people
living below the internationally defined poverty line has
dropped by more than 850 million over the past forty years.
This is an astonishing achievement in itself – all the more
significant as it represents around 75 per cent of the total
reduction in numbers of poor people around the world
achieved through the UN's Millennium Development Goals.

And therein lies a murky story of how the United Nations
systematically manipulates statistics about global poverty

specifically to reassure us that all is well with the prevailing economic paradigm for addressing poverty. Back in 2015, the United Nations published its final report on progress made through the Millennium Development Goals. It claimed that the poverty rate had been cut in half since 2000. But this assertion is based on the *proportion* of people being lifted up out of poverty, not on *absolute numbers*. The reality is that there are still around 1 billion people living in extreme poverty today – more or less the same number as was the case back in the early 1980s.

According to the World Bank, anyone living on $1.90 a day or less is living in extreme poverty. But there is widespread agreement that $1.90 is a wholly inadequate threshold to ensure basic needs are being met, let alone human dignity protected, with a growing consensus that a figure of $5 a day would be far more realistic. This changes the story dramatically: at that level, an astonishing 4.3 billion people are living in poverty, roughly 57 per cent of the whole of humankind. And if that figure was set at $10 a day (that's still only $3,750 a year), there are then 5.1 billion people living in poverty, roughly 67 per cent of humankind.

How can things still be so bad after decades of aid money going to the world's poorest countries – currently at around $150 billion a year? Unfortunately, the true picture of wealth transfers between rich and poor is very different from what we imagine. At the end of 2016, the US-based think-tank Global Financial Integrity revealed the true picture from 2012: while developing countries received about $2 trillion (primarily in terms of aid, new investments, remittances from people living in the rich world, etc.), about $5 trillion moved the other way, from poor countries to rich countries, in terms of debt repayments, repatriated profits for Western multinationals, and vast amounts of 'capital flight' through transfer pricing, 'leakages' in the balance of payments and so on. That's a $3 trillion reverse flow – twenty-four times the annual aid budget.

This whole story is forensically exposed in Jason Hickel's excellent book *The Divide*. He shows how the statistics about global poverty and hunger have been assiduously massaged by the UN to maintain a 'good news narrative' about poverty. For instance, it's so much more reassuring for people in the West to hear that numbers of people living in hunger have declined from 23 per cent of people in the developing world in 1990 to just 15 per cent in 2015, even though in *absolute* terms, that's still 800 million people. As Jason Hickel puts it:

> The good news narrative serves as a potent political tool. It enjoins us to believe that the global economic system is on the right track. It implies that if we want to eradicate suffering, we should stick with the status quo and refrain from making drastic changes. For anyone who has an interest in maintaining the present order of distribution, the 'good news narrative' is a useful story indeed.

The reality of eradicating extreme poverty is really very different. With today's business-as-usual model of economic development, ensuring that everybody could live on more than $5 a day would take 207 years, and the global economy would need to increase to 170 times its present size. Farewell, planet Earth.

Another myth perpetuated by the development industry and by rich world media is that the gap between the richest countries and the poorest countries, in terms of per capita income, has been narrowing. In fact, inequality between countries rose dramatically over the second half of the twentieth century, although that process has now slowed as countries like India and China continue to achieve very high annual levels of economic growth. Levels of inequality are even worse at the individual level. The richest 10 per cent of adults in the world own more than 85 per cent of global household wealth; the top 1 per cent own 43 per cent of global household wealth.

In 2017, Oxfam released its wealth report showing that the richest eight people own as much as the poorest half of humanity – some 3.6 billion. Between them, Bill Gates, Jeff Bezos and Warren Buffett own more than the poorest 160 million people in the USA. In India, the top 1 per cent of the population owns 73 per cent of net wealth.

How we can go on living with this kind of obscene concentration of wealth remains to me a complete mystery. I was one of those naive enough to believe that the Occupy movement, which exploded onto the streets of London and many US cities in September 2011, would bring home to people the full extent of this particular variety of kleptocratic capitalism, and that the 99 per cent of us who are not beneficiaries of the system would start to demand some truly transformational change. But the impacts of the global crash in 2008 were at their most severe at that stage, media attention rapidly waned, and by mid-2012, Occupy was just a passing memory.

However, today's Climate Emergency is bringing a new dimension to bear on the issue of inequality and the superrich – or the 'ultra-high-net-worth individuals', as they like to be called. There are now more than half a million UHNWIs, a number that increases by more than 10 per cent every year. That figure includes more than 2,700 billionaires, according to the Forbes 'World Billionaires List' in 2021. Between them, they have a collective wealth of more than $35 trillion – by way of contrast, the GDP of the USA is around $21 trillion. According to UBS, only about a third of those UHNWIs have more than 1 per cent of their assets invested ethically. In other words, these are the archetypal planet-destroyers. As George Monbiot puts it: 'Immense wealth translates automatically into immense environmental impacts, regardless of the intentions of those who possess it. The very wealthy, almost as a matter of definition, are committing ecocide: there is neither the physical nor ecological space for everyone to pursue private luxury.'

THE POWER OF THE FOSSIL FUEL INDUSTRY

The fossil fuel industry lies right at the heart of this elite nexus of privilege and super-wealth. It epitomises the power of the incumbency order – individuals, institutions, companies and investors who continue to defend a system that has served them so well over so many decades.

Nothing speaks more powerfully to this incumbency dilemma than the issue of fossil fuel subsidies. Every year since 2009, meetings of both the G7 and the G20 have agreed to phase out these subsidies, but literally *nothing* has happened. It is generally accepted that direct subsidies for coal, oil and gas, all around the world, amount to roughly $500 billion a year. This includes both production subsidies for the industry itself and direct support for consumers by keeping the cost of fuel below market prices. In the US, for instance, roughly

$20 billion a year of taxpayers' money is used directly to support fossil fuel companies, with another $14.5 billion used to lower the price of fuel.

That's crazy enough at a time when we need to be phasing out the use of fossil fuels just as rapidly as possible. If you then take into account the so-called 'externalities' caused by the burning of fossil fuels (impact on human health, environmental pollution, and the massive contribution of greenhouse gas emissions to climate change, for which the fossil fuel companies pay literally nothing), the level of *indirect* subsidy is staggering. It's worth quoting directly from the most recent paper on this from the IMF back in May 2019: 'Globally, subsidies remained large at $4.7 trillion (6.3 percent of global GDP) in 2015, and are projected at $5.2 trillion (6.5 percent of GDP) in 2017. [. . .] Efficient fossil fuel pricing (i.e. without subsidy) in 2015 would have lowered global carbon emissions by 28 percent and fossil fuel air pollution deaths by 46 percent, and increased government revenue by 3.8 percent of GDP.' That's the International Monetary Fund talking, not Greenpeace.

The cover-up by oil companies of the knowledge they had of the science of climate change back in the 1970s is now seen by many as 'the most consequential lie in the history of humankind', to use Bill McKibben's powerful words. But these continuing, staggeringly large hand-outs to fossil fuel companies, by governments using taxpayers' money, at a time when we're trying to reduce our dependency on fossil fuels, are without a doubt the most consequential and most reprehensible abuse of taxpayers' money since governments first started taxing their citizens.

Perhaps the greatest strength of Michael Mann's new book,

The New Climate War: the Fight to Take Back our Planet, is his forensic exposé of the role of Big Oil over the last 40 years – and continuing to this day – in denying and obscuring the science of climate change. No individual climate scientist has had to put up with as much vicious, utterly deceitful vilification since the late 1990s, compelling Mann to become 'a reluctant and involuntary combatant in the climate wars'. His exposé includes detailed references to the knowledge that oil companies had of the inevitable impacts of climate change, and the ways in which they used the same dark arts used by Big Tobacco to amplify their lies:

> Their apparent prominence in the public sphere appears far greater thanks to the megaphone provided by the fossil-fuel-funded climate-change denial machine. The megaphone includes Fox News and the rest of the Murdoch media empire, as well as bot armies that are deployed online to flood our social media with misinformation and disinformation. The collective effect is to make extreme positions appear more popular than they actually are. These efforts provide right-wing politicians with talking points and political cover as they continue to do the bidding of the fossil fuel interests who fund their campaigns.

To that record of decades of criminal misinformation and media manipulation must now be added new insights into the way Big Oil lobbied aggressively against clean air regulations. That is despite there being clear evidence that pollutants from the burning of fossil fuels, particularly the smallest particles known as $PM_{2.5}$, could lodge deep in people's lungs causing significant health impacts. It's not too far-fetched to say that hundreds of millions of people have died since as a direct consequence of the success of those companies in preventing timely and proportionate regulation of their toxic products.

For these companies (backed at every turn by their malign cheerleaders in the American Petroleum Institute), they have clearly not yet killed enough people. They're still hard at work lying about the science of air pollution and were of course cock-a-hoop at having had Donald Trump in the White House for 4 years carrying out their bidding via a tragically enfeebled Environmental Protection Agency.

As if to rub salt into these wounds, many of the biggest oil and gas companies continue to spend millions of dollars lobbying against policy proposals designed to tackle climate change. According to the campaigning NGO InfluenceMap, the five largest publicly owned oil and gas companies spend around $200 million a year on such delaying tactics, with BP (at $53 million) and Shell (at $49 million) the worst, with ExxonMobil, Chevron and Total making up the rest. It's an extraordinary thought that we continue to allow these companies to divert some of the money they get from us (as taxpayers) systematically to undermine our interests, and indeed the interests of society as a whole.

Most people would like to see these companies investing far more of their revenues in renewables, and if you listen to their senior executives or are taken in by various advertising campaigns, you might well imagine that's exactly what they're doing. No such luck – or rather, no such leadership. If you look at net investment in low-carbon alternatives as a percentage of total capital expenditure between 2010 and 2018, Total's figure comes in at 4.3 per cent, BP at 2.3 per cent, Shell at 1.3 per cent, Chevron at 0.23 per cent and ExxonMobil at 0.22 per cent. For some, it's too small a figure even to mention in their annual accounts. In 2018, that was an average of just 1 per cent according to the analysis done by Reuters.

This is now changing (with new commitments from both BP and Shell regarding investment in renewables), but it's still slow. Back in Chapter 5, I alluded to the very real possibility of investors temporarily hanging in there with their investments in oil and gas companies until that moment when they suddenly realise the game's up, at which point they all scramble to get out at more or less the same time, causing massive damage to the global economy through what will no doubt be described as this 'unpredictable bursting of the carbon bubble'. 'Unpredictable', perhaps, if you want the exact moment when this will happen, but highly predictable, in fact *guaranteed*, at some point within the next decade. Paul Gilding, one of Australia's leading climate commentators, looks to that moment:

> ... the Climate Emergency meets financial contagion. When the global market flips to FOMO [Fear Of Missing Out] – from fear of acting too early, to fear of being left behind as everyone races for the exits. [...] The financial logic of acting is now impeccable, meaning the only thing left is for there to be a shift in *sentiment* – that moment of an intangible, hard to define flip in how the decision makers in the market see the world. That can happen overnight. And because markets hunt in packs – when they go, they'll all go. [...] The key issue that will drive this loss of value

is not the level of demand *today*, but the level of *belief* that demand will be there in 10–20 years' time. [...] That future demand assumption is based on a belief that climate change is a 'future' risk, that policy is not imminent, that the public isn't engaged, and that the new technologies aren't ready to scale.

Those assumptions were all true 20 years ago. Now they are all wrong. [...] If that sentiment turns, the value is gone. The already struggling oil and gas majors will not then transform; they will just fail, as incumbents usually do.

Oil and gas companies have spent the last decade trying to distance themselves from coal companies, endlessly reminding people that their products are 'cleaner' in terms of emissions of greenhouse gases. But they would be well-advised to analyse the accelerating demise of coal as an object lesson for their own industries. As of the end of 2019, all major insurance companies apart from Lloyd's of London and a few Asian insurers had withdrawn insurance cover for new coal projects – no insurance, no project.

Costs and benefits

There are enormous risks associated with mismanaging the transition away from fossil fuels. But the longer we delay, and the less honest we are about the non-negotiability of that transition, the higher the costs will be. And however substantial those costs may be, even in the most well-managed of transitions, they pale into insignificance when contrasted with the costs of *not* expediting this transition. Back in 2006, Nicholas

Stern's 'Review of the Economics of Climate Change' definitively demonstrated how the benefits of strong early action far outweigh the costs of not acting – and everything that has happened since then has powerfully reinforced that case.

Every now and then, the United Nations Environment Programme produces a Global Environment Outlook (GEO), including details about the number of people affected by environmental disasters. The 2019 GEO pointed to a number of different factors having an ever-worsening impact on the lives of more and more people, with climate change as the biggest 'risk multiplier' of all. In 2018, environmental disasters, including droughts, wildfires, flooding and hurricanes, impacted the lives of 800 million people – more than 10 per cent of humankind. When UNEP first issued its GEO back in 1980, that figure was just 200 million people. Do your own extrapolation.

For that reason, I'm always a little bit sceptical when I hear of the latest finely honed calculation of what the cost to the global economy might be of a 1.5°C, 2°C, or 3°C rise in average global temperatures by 2100. Whatever it is, it will be massively damaging from a macroeconomic point of view, and completely devastating for the lives of billions of people. I think that was the point that David Wallace-Wells was trying to make in *The Uninhabitable Earth* when he speculated that if the average temperature increase by 2100 topped out at 4°C (still perfectly possible unless we radically change our ways), costs could rise as high as $600 trillion – roughly double the total wealth that exists in the world today.

It's the insurance companies that are already picking up the tab here. In 2019, Munich Re (the world's largest reinsurance company) reported that 2017/2018 had been the worst two-year period on record for disasters 'amplified by climate change', with insured losses of $225 billion. Uninsured losses (when costs are borne by individual victims or by governments forced to step in after the event – taxpayers in the US pay nearly ten times more for federal disaster relief than they did back in 1990) come to roughly the same amount. There's something of a myth that the lion's share of the cost of climate disasters is borne by poorer countries; last year, however, Morgan Stanley reported that weather disasters 'intensified by climate change' cost North America $650 billion over the preceding three years – that's two-thirds of all the financial damages suffered worldwide.

Thankfully, 2020 was 'a better year', according to Swiss Re, with insured losses of only $83 billion – still the fifth costliest year on record. No one any longer doubts that both the increased intensity and the frequency of hurricanes and typhoons around the world are caused by climate change (primarily as a consequence of warmer surface temperatures in the world's oceans, with more heat and more water evaporating into the atmosphere), and the same kind of 'attribution science' (see page 88) is now being applied to wildfires. In March 2019 Munich Re explicitly blamed climate change for the $24 billion of losses from the 2019 Californian wildfires. As Ernst Rauch, Munich Re's chief climatologist, said at the time: 'If the risk from wildfires, flooding, storms or hail is increasing, then the only sustainable option we have is to adjust our prices accordingly. In the long run, it might become a social issue. Affordability is so critical, because some people on low and average incomes in some regions will no longer be able to buy insurance.'

'No longer able to buy insurance.' The implications of this are deeply disturbing. The direct costs of Australia's bushfires in 2020 ran to tens of billions of dollars; the longer-term health costs, from air pollution, could be equally onerous. And at what point will Australia's insurers either red-line heavily forested areas as 'uninsurable', or jack up premiums so high that living there becomes unaffordable?

Facing incalculable economic dislocation on that scale means we have to be prepared to invest *now* to avoid such a prospect. When the UK became the first major economy in June 2019 to commit to a Net Zero economy by 2050, the positive impact of this announcement was immediately undermined by the Chancellor of the Exchequer at that time warning that it would 'cost £1 trillion' to achieve such a goal, without a passing reference to any benefits, let alone to the climate imperative. Rather than fear-mongering about astronomical *costs*, the UK's Committee on Climate Change has estimated it could require *investment* of between 1 and 2 per cent of GDP every year in order to get to a Net Zero economy by 2050. True enough, to get there faster will undoubtedly cost more; the Green Party in the UK is pressing for a 2030 deadline, which would require an investment plan of something much closer to 4.5 per cent of GDP. Frankly, there is no pain-free route to a Net Zero economy, but the longer we prevaricate, the greater the pain will be.

LEARNING FROM HISTORY

Ironically, we've been here before – more than thirty years ago. When I was director of Friends of the Earth back in the late 1980s, our membership recruitment was greatly bolstered by Prime Minister Margaret Thatcher, a chemist by training, suddenly becoming aware of the threat posed to the future of humankind by the build-up of greenhouse gases in the atmosphere. It became one of her top priorities; she summoned her entire Cabinet to a day-long seminar on climate change in Downing Street, and delivered a series of quite remarkable speeches, including an impassioned plea to the United Nations General Assembly in November 1989. Unfortunately, her fellow world leaders were largely unmoved; there was no real follow-up, and Mrs Thatcher had been forced to resign a year later in November 1990. She herself took no further part in the climate change debate.

When the Thatcher archives from 1989 were released in 2019, they included a draft of an earlier speech she gave to the Royal Society in 1988, which included the following paragraph:

> We may need to take major steps, perhaps including a world levy on fuel prices to support improved energy efficiency. Part of the solution to deforestation might be to link the protection of trees in third world countries to debt retirement.

By any standards, that is quite astonishing: there's Mrs Thatcher, one of the world's most powerful and admired leaders, telling her colleagues more than thirty years ago that they would need to accept some kind of levy on fossil fuels if the looming threat of global warming (as climate change was referred to back then) was to be dealt with. Just imagine for a moment if everybody had sat up, paid attention, and got on with introducing such a transformative policy through the UN's Framework Convention on Climate Change when it was brought into being three years later in 1992. We would not have wasted these thirty precious years – well over half the greenhouse gases in the atmosphere have been released since 1988.

But that paragraph of Mrs Thatcher's did not survive into the final draft of the speech. It was axed at the behest of Nigel Lawson, then Chancellor of the Exchequer, with the following comment from his private secretary: 'The Chancellor has seen the draft speech. His main comment is that the paragraph beginning "We may need to take major steps . . ." *must* be deleted. These bizarre ideas are contrary to Government policy, and political dynamite.'

It's hard to exaggerate the significance of this tiny redrafting moment. In those few words you see one of the earliest signals of what became a thirty-year campaign by climate denialists and free-market zealots to ensure that as little as possible was done to address the challenge of climate change. Mrs Thatcher may have rapidly bowed out of the climate debate, but Nigel Lawson did not. In 2009, he set up the Global Warming Policy Foundation, one of the most persistent and influential of all the denialist think-tanks on either side of the Atlantic.

This chapter is primarily about the power of the fossil fuel incumbency, and its malign and deeply reprehensible role in blocking effective climate policy. The industry has been at the heart of the neoliberal, free-market ideology that has dominated the global economy since the late 1980s. Even then, leading advocates of this kind of laissez-faire capitalism realised that dealing with climate change would be highly problematic from their perspective. For one thing, countries would have to work together as collaboratively as possible, even 'pooling sovereignty' (as happens in determining climate policy within the EU) rather than promoting competition at every turn. It was clear that markets alone would never be able to come up with appropriate responses to climate change – which meant governments would have to intervene substantively, using fiscal policy, rigorous regulation and standard-setting, incentivisation schemes and so on – 'big government' in other words, which is total anathema to these free-market zealots. As Naomi Klein has pointed out:

> The late 1980s was the absolute zenith of the neoliberal crusade, a moment of peak ideological ascendancy for the economic and social project that deliberately set out to vilify collective action in the name of liberating free markets in every aspect of life [. . .] Just as governments were getting together to get serious about reining in the fossil fuel sector, the global neoliberal revolution went supernova, and that project of economic and social reengineering clashed with the imperatives of both climate science and corporate regulation at every turn.

That's what killed off those early efforts to get to grips with climate change in the late 1980s and 1990s – not ignorance, or indifference, or selfishness, or greed, or any other aspect of 'human nature' that has been summoned up as a possible explanation for these three wasted decades. Nathaniel Rich wrote an entire book (*Losing Earth: The Decade We Could Have Stopped Climate Change*) blaming ordinary people for this historical failure at the end of the 1980s: 'The conditions for success could not have been more favorable. All the facts were known, and nothing stood in our way. Nothing, that is, except ourselves.' Wrong, wrong, wrong: what really stood in our way then, and still does today, is a powerful, ruthless, self-serving political elite that will brook no barriers to further enriching itself at the expense of the whole of the rest of humankind.

So here's the rub for those who still hope that we can avoid any kind of political confrontation, that we can do everything we need to on a bipartisan basis, that technology alone will get things sorted: regrettably, today's unreconstructed neo-liberal capitalism still rules the roost in the USA, the UK, Australia, Brazil and many other countries, as well as in critical institutions like the IMF, the WTO and so on. Despite all its calamitous failures over the past thirty years, its hold on the levers of power is as strong as ever; it's not going anywhere, anytime soon.

Standing up against that powerful elite is therefore an essential part of today's climate activism, whatever one's own personal political affiliation. First and foremost, that means protecting the democratic freedom on which all such political activity depends.

16

DEMOCRACY AT RISK

'Recognising that sustainable development,
democracy and peace are indivisible is an idea whose
time has come.'

WANGARI MAATHAI

In 2010, a decision in the US Supreme Court (the 'Citizens United' judgment), removed virtually all limits on contributions to political campaigns, and confirmed that all such contributions could be made anonymously. Many have suggested that the principal consequence of this has been to make the USA one of the most legally corrupt nations on Earth. Countless billions of dollars have flowed from an undeclared number of those US-based HNWIs I referred to in the preceding chapter into supporting right-wing, neoliberal candidates at every level of US democracy – from the local school board to the presidency itself.

It's obviously very hard to unearth the full extent of this corrupting influence, but Jane Mayer's *Dark Money* (published in 2016) remains one of the most authoritative accounts I've ever read. She focuses much of her analysis on the Koch brothers, Charles and David, billionaire proprietors of an oil company based in Kansas, and since the 1980s the most generous of all the funders of individuals, institutions and organisations subscribing to their extreme free-market and climate-denying views. If you had to name two individuals who have done more to exacerbate the existential risk that we are now facing through accelerating climate change, it would undoubtedly be the Koch brothers.

But this kind of dark money, moving invisibly and unchecked from account to account, is of course a global phenomenon. A network of global tax havens makes it possible for all rich people, not just the ultra-high-net-worth individuals, to minimise, avoid and even evade paying their taxes – the first two strategies still being wholly legal. The UK is right at the heart of this particular spider's web, with tax havens in the British Virgin Islands, Bermuda, the Cayman Islands, Turks and Caicos, Anguilla, the Isle of Man, Jersey and Guernsey.

In May 2019, the latest report from the Tax Justice Network demonstrated that 'The UK with its corporate tax haven network is by far the world's greatest enabler of corporate tax avoidance, and has single-handedly done the most to break down the global corporate tax system, accounting for more than a third of the world's corporate tax avoidance risks [. . .]. That's four times more than the next greatest contributor of corporate tax avoidance risks, the Netherlands, which accounts for less than 7 per cent.' It's reckoned that somewhere between $500 billion and $600 billion is lost to governments every year in corporate tax revenues through legal and sort-of legal schemes. Beyond that, the Tax Justice Network estimates that a further \$21–32 trillion of private wealth is held offshore, meaning that governments are losing out to the tune of at least another \$200 billion a year via hidden offshore wealth.

It's madness. With governments endlessly complaining about trying to 'balance the books', often imposing the cruellest of austerity measures on their own long-suffering people, why do we allow the already super-rich to escape from one of the basic obligations of citizenship: to contribute to a nation's wealth via fair and proportionate taxation? The answer is self-evident: many politicians are themselves beneficiaries of these deeply unethical practices. When the Panama Papers were leaked in April 2016, highlighting examples of tax haven abuse, twelve national leaders and 143 leading politicians were implicated in the revelations. As Oliver Bullough points out in *Moneyland*:

> Money flows across frontiers, but laws do not. The very wealthiest people have tunnelled into this new land that lies between all our nation states, where borders have vanished. They move their money and themselves wherever they wish, picking and choosing which countries' laws they wish to live by. Until this situation changes, the lawyers, bankers and drafters of deliberately loose financial laws will continue to prosper.

There are many hidden costs here too, in terms of the erosion of trust that people have in their democratically elected politicians – particularly in the US and the UK. People read about these scandals; they know they're pervasive; they know, deep down, the system isn't 'fair', and that they're the ones paying the price in terms of failing public services, closed libraries, disintegrating infrastructure, neglected public spaces and a general collapse in civic values. According to the study done for the World Values Survey in 2016, there are now significant minorities in both the USA and Europe who see 'having a democratic political system as a bad or a very bad way to run this country'. The figures have been rising steadily since the mid-1990s: the percentage of US respondents in 1995 was 17 per cent; by 2011, it had risen to an astonishing 24 per cent. Even in some of the richest and most politically stable regions of the world, it seems as though democracy is in a state of serious disrepair.

This has to be seen as something of a historical tragedy. In 1940, there were no more than a dozen fully functioning independent democracies in the world; by 2000, only eight nation states had never had any kind of democratic election. That's a remarkable pro-democracy surge over sixty years. But twenty years on, it's a very different picture. Every year, *The Economist* magazine's Intelligence Unit publishes its Democracy Index. In January 2020, it concluded that only 430 million people, in 22 countries, are living in a 'full democracy'.

That grim assessment is confirmed by the democracy watchdog V-Dem, which calculates that almost a third of the world's people who now live in notionally democratic countries are experiencing what is classified as 'autocratisation', where some of the most critical elements of a functioning democracy (such as free and fair elections, freedom of assembly, an independent judiciary and a free press) have already been significantly weakened or undermined. That's the case, according to V-Dem, in twenty-four different democracies today, including Brazil, India and the United States – where what is often described by Democrats as 'a slow twenty-five-year march to the right' has led to widespread voter suppression, persistent gerrymandering (manipulating the boundaries of voting districts to favour Republican candidates), institutionalised racism in voter registration processes, and so on.

Even in the UK (still categorised as a 'full democracy' in the Democracy Index), the Hansard Society's annual audit in April 2019 showed anti-democratic sentiments growing rapidly. Asked if they agreed that 'Britain needs a strong ruler willing to break the rules', 54 per cent of respondents agreed that we did, more than twice as many as those who said that we didn't. In Boris Johnson, the UK has certainly got one of those rule-breaking rulers, so we'll just have to see what this means for our democracy.

THE POWER OF POPULISM

Much of this is down to the rapid growth of populist parties of one kind or another in many different parts of the world. Some interesting research by psychologist Karen Stenner in 2007 suggested that around a third of people in any country are psychologically predisposed to favour 'oneness and sameness' over 'freedom and diversity'. Once persuaded that their culture is under siege and could even be destroyed (under challenge from minorities or migrants, for instance), these people can be easily persuaded by unscrupulous populist politicians to support more authoritarian policies. In Europe, Viktor Orbán in Hungary (theoretically a thriving democracy; in reality, a one-party state) and Matteo Salvini in Italy have very successfully manipulated Europe's ongoing refugee crisis.

Salvini's League is currently polling at around 25 per cent, and is still the largest party in Italy, systematically sowing division, stoking resentment and exploiting prejudice.

But for every social movement there are invariably counter-movements, and the eruption in November 2019 of the 'Sardines', filling squares and piazzas across Italy to protest against Salvini's 'hateful language and vulgar politics', has been remarkable. We saw the same phenomenon in the election of Ekrem İmamoğlu as mayor of Istanbul in June 2019, with a campaign based on tolerance and overcoming division, specifically taking on the populist, hate-driven politics of Turkey's strongman president, Recep Tayyip Erdoğan – rejecting his 'climate of fear' with a 'climate of love'.

One of the hallmarks of today's populism is its direct appeal to those who feel themselves to have been 'left behind' by forty years of globalisation, betrayed by governments made up of aloof, self-serving elites indifferent to their lack of opportunity and chronic economic insecurity. The prospect of near-permanent unemployment, or 'underemployment' in the gig economy, crushes individual and societal hope; constant social inequality is the norm for so many in today's new economy. This is particularly true among younger people, and especially men; 60 per cent of Jair Bolsonaro's supporters in Brazil are aged under thirty-four.

And there are more profound consequences that have to be addressed, in terms of the impact of populism on our overall capacity, as a family of nations, to address today's converging crises. In a recent article, Paul Ehrlich and colleagues warned of the potential significance of populists successfully 'weaponising' environmental concerns:

The continued rise of extreme ideologies is likely, which in turn limits the capacity of making prudent, long-term decisions, thus potentially accelerating a vicious cycle of global ecological deterioration and its penalties. This includes the weaponisation of 'environmentalism' as a political ideology, rather than it being viewed as a universal mode of self-preservation and planetary protection that ought to transcend political tribalism. Indeed, environmental protest groups are being labelled as 'terrorists' in many countries.

But it is surely the cruellest irony of our age that those who have proved to be most effective in manipulating this wholly understandable anger and disaffection are themselves powerful embodiments of today's callous, super-rich elite. To hear politicians like Donald Trump and Boris Johnson speak on behalf of the left-behind in both the US and the UK makes my blood boil, and the unending spate of lies necessary to sustain that uncaring hypocrisy has coarsened politics across the Western world.

Donald Trump appeared to revel in his role as 'coarsener-in-chief', 'a larger-than-life, over-the-top avatar of narcissism, mendacity, ignorance, prejudice, boorishness, demagoguery and tyrannical impulses', as Michiko Kakutani so memorably described him in her deeply disturbing book, *The Death of Truth*. 'Trump is a troll – both by temperament and by habit. His tweets and off-hand taunts are the very essence of trolling – the lies, the scorn, the invective, the trash talk, and the rabid non-sequiturs of an angry, aggrieved, isolated and deeply self-absorbed adolescent who lives in a self-constructed bubble.'

It was hard not to be fascinated by such an outlandish caricature of a person, but this should in no way obscure the incredibly serious consequences of his unremitting, day-by-day assault on truth, on the American constitution, and on the rule of law – as was so painfully exposed in the investigations into his 'improperly' pressuring the Ukrainian government to dig up damaging information on Joe Biden, selected as the Democratic candidate in the 2020 presidential election. This investigation led to his impeachment in December 2019, only for him to be found not guilty of 'high crimes and misdemeanours' by his 'peers' in the US Senate in February 2020 – a decision that conclusively confirmed the demise of the Republican Party as a moderate conservative force in US politics. This is another hallmark of contemporary populism: you are either 'with the project' or 'against the project', with large numbers of people once thought of as a stabilising force in society becoming increasingly polarised. Moderate conservatives are forced either into resentful disengagement or ever more hard-line positions.

In their important book *How Democracies Die*, political scientists Steven Levitsky and Daniel Ziblatt show how this process has played out in countries like Turkey, Hungary, Poland and Russia, as the 'guardrails of democracy' are progressively removed, and those institutions once established to serve and protect the public become compliant extensions of the ruling party. This has undoubtedly happened in the USA, as borne out by the Republican Party's apparent indifference to the wholesale interference by Russia in the 2016 presidential election. As confirmed by the US Director of National Intelligence in January 2017, President Putin's

Internet Research Agency created thousands of fake social media accounts to boost Donald Trump's candidacy, reaching tens of millions of social media users between 2013 and 2017 with fabricated articles and a non-stop barrage of fake news – the so-called 'firehose of falsehood'. This is not disputed (notwithstanding Trump himself dismissing the whole thing as 'a hoax'), even if the subsequent investigation by Robert Mueller came to the conclusion in April 2019 that there was insufficient evidence to bring any charges of conspiracy or direct co-ordination against the president himself.

In the end, America's idiosyncratic democracy withstood the four years of constant undermining by Donald Trump. But the shameful events of 6 January 2021, when a mob of supporters of Donald Trump set out to occupy Capitol Hill to prevent the ratification of the votes from the 2020 Presidential Election, will resonate in US politics for a very long time to come. He was impeached for a second time for directly instigating that riot, which caused the death of five people, but there weren't enough votes in the Senate for his indictment to carry. Seven Republican Senators voted with the Democrats to impeach him, but the rest chose to maintain their support for what is now known as 'the Big Lie' – Trump's persistent claim that the election was rigged and that he won by a landslide.

The one consolation emerging from this trauma for the American people is the growing pressure on the Republican Party to repudiate its allegiance to Trump, and to redis-cover what it once stood for. Given that a large number of Republicans have already decided to stick with Trump and his Big Lie, at least until the mid-term elections in 2022, and a rather smaller number have decided that they will never have anything to do with him again, there is a very real possibility that the Republican Party will split. US democracy would be a clear winner from such a momentous schism.

MANIPULATING THE MASSES

Security services the world over have had to accept that some of the most dangerous threats to national security today are no longer physical but entirely digital, with rapidly evolving technology (including artificial intelligence) allowing both hostile nations and co-ordinated groups to step right into the heart of any country's democracy with a powerful online presence. This poses an ever-greater threat to even the most secure of democracies, and the scale of it is already astonishing. In the run-up to the European elections in May 2019, the UK's Institute for Strategic Dialogue unearthed a wide range of platforms seeking to promote right-wing, antisemitic and racist positions:

> The team estimates that far-right disinformation networks across France, the UK, Germany, Spain, Italy and Poland produced content that was viewed an astonishing 750 million times in three months. In Poland, pro-government accounts posed as pensioners in order to attack striking teachers [...] A network of 60 pages on Facebook also amplified antisemitic and pro-Kremlin content. In Germany, 200,000 fake social media accounts were spreading electoral content supported by the far-right political party Alternative für Deutschland. In Italy,

a network with more than 2.6 million followers spread anti-migration, antisemitic and anti-vaccine information. An estimated 9.6 million Spanish voters had seen disinformation on WhatsApp.

Facebook was notified of 500 different hate sites spreading this poison; by the time of the election, it had taken down just seventy-seven of them. Earlier in the year, Facebook itself announced that it had identified a network of 137 accounts engaged in what it called 'inauthentic activity' targeting the UK with similar material, pumping polarised messages to both ends of the political spectrum in order to whip up outrage. Facebook was not able to say who was responsible, and obstinately continues to defend its wholly inadequate strategy of seeking to reduce misinformation and foreign interference by removing fake accounts. It simultaneously announced that it would no longer be fact-checking adverts posted directly by politicians and their campaigns. 'We do not believe it should be our role to fact-check or judge the veracity of what politicians say,' said Richard Allan, Facebook's vice president of policy solutions.

By contrast, Jack Dorsey, Twitter's CEO, decided at the same time to stop all political advertising on Twitter globally. 'While internet advertising is incredibly powerful and very effective for commercial advertisers, that power brings significant risks to politics where it can be used to influence votes to affect the lives of millions.'

The failure to properly regulate, let alone properly tax, these mega-monopolies, now in full-on robber baron mode, is one of the most egregious collective failures of modern government. But bit by bit, today's tech platforms (Facebook (including Instagram and WhatsApp), Twitter, Google (and its owner Alphabet) and Amazon) are coming under growing scrutiny, with more and more powerful voices pressing not only for more stringent regulation, but for breaking up such unaccountable giants. In October 2020, the US Department of Justice filed an anti-trust case against Google, seeking to put an end to all anti-competitive contractual conditions mandating the use of Google's search engine. Elsewhere, Senator Elizabeth Warren has consistently raised concerns about Facebook, which controls about 80 per cent of global social networking revenue: 'Facebook has too much power over our economy, our society and our democracy.' Mark Zuckerberg, Facebook's CEO, has proved to be particularly obtuse, refusing to contemplate even quite minor governance reforms to limit his near-absolute power within the company, and alienating policymakers on both sides of the Atlantic with his unique variety of passive-aggressive arrogance.

The tech giants are now fighting hostile battles on many different fronts. Their unique ability to avoid taxes has outraged more and more politicians. The UK-based NGO Fair Tax Mark reckons that the big six US tech companies (the four above plus Microsoft and Netflix) have 'aggressively avoided' $100 billion of tax in the UK alone over the past ten years. And the confirmation in December 2019 of the formidable Margrethe Vestager as the EU Commission's 'executive vice president: a Europe fit for the digital age' puts them all on notice that she'll be clamping down on their more outrageous monopolistic and tax-avoiding abuses.

Nor can these companies avoid their share of responsibility

for some of the more profound impacts of the internet on society. Will Storr's *Selfie: How the West Became Self-Obsessed* highlights soaring rates of self-harm, eating disorders, anxiety and body dysmorphia among young people, with more and more feeling lonelier and increasingly isolated. From time to time, stories of extreme individual pain, caused by online bullying, harassment or more subtle 'peer group pressure' break through into the media; the occasional suicide tears at our heartstrings. But nothing much happens. Meanwhile, the collective, societal cost of worsening mental health among the young becomes 'just one more problem' to be dealt with by our massively overstretched health services.

DEMOCRACY FIRST

All these different factors (dark money, populism, the 'barbarisation' of politics, digital interference in election campaigns, the assault on US democracy, the unaccountable power of big tech, the damage done through the internet to the lives of so many young people, and so on) are having a pronounced impact on the state of democracy today. Yet we seem to go on ignoring all these flashing signals. Back in September 2018, Martin Wolf of the *Financial Times* warned of the consequences of there having been so little change in both the US and Europe since the financial crash in 2008: 'If those who believe in the market economy and liberal democracy do not come up with superior policies, demagogues will sweep them away. A better version of the pre-2008 world will just not do. People do not want a better past; they want a better future.'

As I've said, however, the signs are not good. The 2016

World Values Survey I referred to earlier showed that the younger people are, in both the US and Europe, the less likely they are to care about the state of democracy. Rising support for illiberal politics is growing fastest among the rich elite in the USA, and this is particularly true among well-off young people: asked in a poll in 2019 if it would be 'a good thing for the army to take over', 35 per cent of relatively well-off young Americans agreed that it would be; that figure was just 8 per cent in 1995.

President Biden is acutely aware of this worrying trend and takes every opportunity he can to affirm the centrality of strong democratic institutions in our prospects for a better world. These were his comments to the Munich Security Conference in February 2021:

> We're at an inflection point between those who argue, given all the challenges we face – from the Fourth Industrial Revolution to a global pandemic – that autocracy is the best way forward, and those who understand that democracy is essential to meeting those challenges. We must demonstrate that democracies can still deliver for our people in this changed world. Democracy doesn't happen by accident. We have to defend it, fight for it, renew it. We have to prove that our model isn't a relic of history; it's the single best way to revitalise the promise of our future.

His first line of defence will need to be in the USA itself. A clear majority of Republicans in both the House of Representatives and in the Senate are still supporting the 'Big Lie' – that Donald Trump won the 2020 Presidential Election by a landslide only to have it 'stolen' from him by corrupt officials and politicians at state level. Since that election, Republicans in 43 states have proposed new laws to make it even harder for citizens to vote (where voter suppression is already a significant problem), with a particular push in Georgia, Arizona, Michigan and Pennsylvania – some of the most closely fought states in the 2020 election.

For many commentators, this makes the 'For the People Act' (now reintroduced in Congress), which will expand voter rights, limit the gerrymandering of electoral boundaries and change campaign finance laws to curtail the influence of big money – the single most important legislative initiative of the Biden–Harris administration.

Back in 1951, in her *The Origins of Totalitarianism*, Hannah Arendt wrote that 'The ideal subject of totalitarian rule is not the convinced Nazi or the convinced Communist, but people for whom the distinction between fact and fiction (i.e. the reality of experience) and the distinction between true and false (i.e. the standards of thought) no longer exist.' Even as we celebrate the incredible power of the new-style leadership of Greta Thunberg, Alexandria Ocasio-Cortez and so many other young champions of radical climate action, it's uncomfortable to have to recognise these anti-democratic trends not just among older people, but among some of today's younger generation – particularly among those who feel their life chances have already been significantly diminished by years of austerity and economic neglect.

If our democracies are already struggling to reconcile the conflicting demands of all today's multiple interest groups and political persuasions, just imagine how hard it will be for those democracies to start incorporating the interests (the *rights*, even?) of future generations. In Chapter 5, I wrote about various young people's lawsuits, raising difficult questions about intergenerational responsibilities and ethical duties across decades. We're only just coming to terms with the fact that even if we succeed in dramatically decarbonising our economies, sea levels will go on rising and climate shocks intensifying for many, many years to come. Astra Taylor explores these time-bound dilemmas in her new book, *Democracy May Not Exist, But We'll Miss It When It's Gone*:

> We are all born into a world we did not make, subject to customs and conditions established by prior generations, and then we leave a legacy for others to inherit. The project of self-government invariably requires navigating the tension between short- and long-term thinking [. . .] nothing illustrates this more profoundly than the problem of the climate crisis, which calls into question the very future of a habitable planet. Our democratic movements must be guided by a deceptively simple question: what kind of ancestors do we want to be?

Even in the short term, the demands of dealing with accelerating climate change will pose some extremely difficult challenges for our already weakened and self-doubting democracies. As we saw in Chapter 12, buried deep in the debate about the pros and cons of different kinds of geoengineering is the lurking assumption that we may need to 'impose emergency measures' if we fail to decarbonise fast enough – including schemes to inject huge volumes of aerosols into the upper atmosphere to reflect back incoming solar radiation, however disturbing the risks might be.

It's been clear to me for a long time that defending democracy is an integral part of the work we need to do as climate activists. In my darker moments, I find it all but impossible to disconnect the threat of runaway climate change from the threat of runaway neo-fascism. And those moments don't get any brighter when I think about the dire state of the natural world.

17

PLANETARY PRESSURES
AND OPPORTUNITIES

'Those who contemplate the beauty of the Earth
find reserves of strength that will endure as long as
life lasts.'

RACHEL CARSON

'The health of the ecosystems on which we and other species depend is deteriorating more rapidly than ever. We are
eroding the very foundations of economies, livelihoods, food
security, health and quality of life worldwide.'

This is the headline conclusion from Bob Watson, chair and
lead author of a landmark report published in May 2019 for the
Intergovernmental Science-Policy Platform on Biodiversity
and Ecosystem Services (IPBES). Clumsy name; devastating
report. Compiled by 145 expert authors from fifty different
countries, reviewing more than 15,000 scientific papers and
government sources, it leaves literally no room for the remotest
lingering doubt, concluding its analysis with the sledgeham-
mer statistic that at least *1 million species* are now at risk of

extinction. This 'sixth mass extinction' has been talked about for a long time. The previous five were all caused by natural phenomena such as meteorite impacts or volcanic eruptions, but this one is 100 per cent down to us. Which is why some call it a 'mass extermination' rather than a mass extinction.

The IPBES is the equivalent in the world of biodiversity of the IPCC in the world of climate change. And in the same way that the IPCC is now commenting much more sharply on an indisputable acceleration in a host of different climate impacts, so the IPBES has described the acceleration in impacts on the natural world as 'unprecedented'. People are increasingly beginning to realise that we have to think about accelerating climate change and biodiversity loss as mutually reinforcing emergencies.

This was disturbingly confirmed at the UN climate conference in Madrid at the end of 2019, with a new report from the International Union for Conservation of Nature highlighting the growing problem of 'de-oxygenation' in the world's oceans. As the oceans get warmer (heat in the world's oceans reached record levels in 2019 as a direct consequence of climate change), its waters hold less oxygen, and there's less mixing up of oxygen-rich surface water with deeper waters which hold less oxygen. This effect is compounded by more and more coastal areas being increasingly affected by run-off of nutrients from intensive agriculture, creating 'dead zones' all around the world.

There are now more than 700 sites worldwide affected by low oxygen conditions (up from just forty-five in the 1960s), already impacting many species. Because of their size and energy demand, species such as tuna, marlin and sharks need oxygen-rich water, which means they're being driven into waters nearer the surface, making them more vulnerable to over-fishing. De-oxygenation is also starting to alter the balance of marine life, favouring low-oxygen species (such as jellyfish and squid) at the expense of many fish species.

These climate-biodiversity synergies have started cropping up in the scientific record with depressing regularity – and 2019 was the year when the fate of the world's insects shot to the top of the agenda as the most startling example of that. From the point of view of today's sixth mass extinction, insects are right up there as the worst affected of all the different categories of species. A report published in *Biological Conservation* back in February 2019 showed that the rate of extinction in insect species is eight times faster than that of mammals, birds and reptiles. The total mass of insects is falling by a precipitous 2.5 per cent a year; a third of all insects are now endangered.

Insects are fundamental to the whole of life on Earth, as pollinators (80 per cent of wild plants are estimated to depend on insects for pollination), as food for other creatures (60 per cent of birds rely on insects as a food source), and as recyclers of other creatures' dung (think 6,000 different species of dung beetle). The report (now colloquially referred to as the 'Insectageddon' report!) was also uncharacteristically outspoken for a peer-reviewed study of this kind, directly pointing the principal finger of blame at modern agriculture and the excessive use of pesticides, while acknowledging that accelerating climate change is also beginning to have an ever more serious contributory impact.

Modern intensive farming is an extraordinarily toxic killing machine. The agri-chemical companies would have us believe that they're responding to this crisis – by seeking to find ways of maximising the lethality for the particular pest that they're trying to take out, while minimising the 'collateral damage' done to other living organisms. But it simply isn't working. Nearly sixty years on from Rachel Carson's *Silent Spring*, it's clear that the only conceivable rationale for perpetuating this war on nature is the pursuit of profit itself, rather than the optimisation of environment-friendly farming practices. Constant applications of pesticides and fertilisers permanently disrupt the balance of microorganisms in the soil, fundamentally undermining one of the most important biological partnerships powering all life on Earth – between plants and the mycorrhizae on which they depend. Mycorrhizae (from the Greek words '*myco*' for fungi, and '*rhizae*' for roots) provide nutrients for the plants (nitrogen, phosphorus, etc.), and in return the plants provide carbon-based sugars (often referred to as 'exudates') for the bacteria and fungi.

Six decades of chemical farming has screwed all that up good and proper. The state of the world's soils is dire. The Food and Agriculture Organization has acknowledged that as much as 25 per cent of the Earth's productive land is now so degraded that it can no longer be used, and that at least 24 billion tonnes of fertile topsoil are lost every year to erosion, deforestation and wholly unsustainable chemical-dependent farming practices. Soil depth in Iowa (at the very heart of the US breadbasket) has gone from 9.5 inches (24cm) in 1950 to 4.8 inches (12cm) in 2017. A report in 2015 from the think-tank Economics of Land Degradation estimated that degradation of one kind or another costs the global economy between $6.3 trillion and $10.6 trillion every year.

And soil erosion is going to get a lot worse – as a consequence of dramatic changes in rainfall patterns as a result of climate change. One of the most widely accepted findings in climate science is that higher temperatures mean more water vapour in the atmosphere, which increases the incidence of heavy downpours of rain – sometimes described as 'rain bombs'. These downpours have a much more damaging impact on soils than regular rain, often causing extremely serious erosion. This will have a major impact on annual productivity growth in corn, wheat, soy and rice yields, at what is already a worrying time for crop yields. At the time of the Green Revolution in the 1960s and 1970s, yields were increasing by around 3.5 per cent per annum, as a result of genetics and better agronomy; that's now down to around 1.2 per cent – a figure that has to be set against the annual rate of growth in the global population, which now stands at 1.05 per cent.

The FAO now recognises that soil degradation is due primarily to the loss of organic matter in the soil, particularly soil carbon, but including the sum of hugely diverse organisms such as fungi, invertebrates such as worms, slugs, grubs and other insects, literally countless microbes, as well as root matter and decomposing vegetation. Soil carbon is considered to be the single most important element in this mix because it sustains huge populations of microorganisms, and determines the availability of nutrients for plant growth. As Kristin Ohlson puts it in her wonderful *The Soil Will Save Us*:

Weirdly, we've all been schooled in the notion that plants are takers, removing nutrients from the soil and leaving it poorer. But when plants are allowed to work with their partners in the soil, they're givers. They feed carbon exudates to the community of bacteria and fungi to keep them thrumming with life and pulling mineral nutrients from the bedrock as well as from particles of sand, silt and clay, because they know – if that word can be applied to organisms without brains – that they will profit from the gift. When predators in the soil eat the bacteria and fungi, all those nutrients are released near the plant. There's always enough, unless humans or some other force messes up the system.

I'll come back to the implications of that little lesson in biochemistry later in the chapter, but I just need to round out this short visit to the world of synergies by stressing the critical links between soil degradation on the one hand and deforestation on the other. The FAO is more and more aware of this, and has shown that one of the reasons for Asia's very high rates of soil degradation has been continuing deforestation, particularly in countries like Indonesia, Thailand, Cambodia, Vietnam and now Myanmar. But this is a global phenomenon. How many more times will we have to read of devastating landslides or mud-slips caused by deforestation in the surrounding hills? Much of that deforestation is caused by people desperate to find enough land for themselves and their families on the edge of existing towns and cities.

In February 2021, Sir Partha Dasgupta published his 'The Economics of Biodiversity', an extraordinary review of all these continuing impacts on the natural world, commissioned by the UK Treasury. It reminds us just how much is at stake right now, and comes up with a wealth of recommendations for consideration, in the first instance, at the critical UN Biodiversity Conference in China in October 2021.

Unlike a lot of reports of this stature, it pulls no punches in identifying population growth as a highly significant driver of continuing environmental destruction ('Growing human populations have significant implications for our demands on nature, including future patterns of global consumption'), and advocates strongly for the global community ramping up investment in family planning and reproductive healthcare.

DOWN TO THE NUMBERS

Complex synergies. Hidden dependencies. Everything is connected to everything else. 'There is only one soil, one flora, one fauna, and one people. Hence only one conservation problem.' Wise words, written eighty-five years ago by the great Aldo Leopold in his lecture on 'Land Pathology'. They remind us that there's one other factor linking all the problems I've touched on in this chapter – in the world's oceans, forests and soils – and that's the sheer number of human beings bringing ever greater pressure to bear on all these complex, interconnected systems.

At one level, the numbers themselves are easy to grasp. The human population is expected to grow to around 11 billion by the end of the century, up from 7.8 billion today – an additional 3.2 billion. Most of that growth will come in Africa, the Middle East and South Asia (Pakistan, Afghanistan and parts of India), where average fertility rates (the number of children a woman has in her lifetime) remain high. In much of the rest of the world, average fertility has fallen significantly, and is now below what is called replacement level at 2.1 children per woman, including all European countries, the US, Canada, Australia, Brazil, Russia, Japan, South Korea, Iran and China. Numbers alone don't tell the full story; levels of per capita consumption are an essential consideration too. Generally speaking, as levels of consumption (or affluence) increase, particularly in rich world countries, so too does the impact on the natural world.

Over the years there have been *many* eloquent appeals from scientists for politicians to focus on the challenges of continuing population growth, including the 2017 Scientists' Warning to Humanity, signed by more than 15,000 scientists from across the world. Among its thirteen recommendations to end the destruction of our natural environment and start managing climate change were reducing fertility rates through education and family planning, as well as through the reduction of poverty and inequality, rallying political leaders behind the goal of establishing a sustainable human population. The warning was updated in 2019, with an even more desperate call for action on these matters.

The biggest challenge of all is population growth in Africa. Fertility rates in Africa are declining only very slowly, and in many countries continuing population growth is actively encouraged for political or cultural reasons. Working off the UN's World Population Prospects (September 2019), the whole of the rest of the world apart from Africa is scheduled to see limited population growth, up from 6.6 billion in 2019 to 7.6 billion by 2050, and then it drops back to 7.1 billion by 2100. By contrast, Africa grows from 1 billion in 2019 to 2.1

billion in 2050, and a staggering 3.7 billion in 2100. Nigeria grows from 200 million today to 400 million by 2050, to 730 million by 2100. Contraceptive use in Nigeria remains stubbornly low, and nearly half the population believe contraception to be either amoral or against their religion.

The self-evident links between continuing population growth, the denial of women's rights and today's climate and ecological emergencies seem utterly obvious. But most environment and development NGOs in the West will still not properly acknowledge that population growth is one of the principal drivers of both emergencies.

There are many reasons for this continuing and increasingly perverse variety of denialism. They fear that the work they do to highlight the appalling impacts of overconsumption in the West will somehow be weakened by talking about overpopulation. They don't want to get caught up in the inevitable controversies surrounding the population debate, including immigration issues. It's easy to be perceived as patronising or insensitive in addressing today's population challenge as a white, Western environmentalist, especially if you're a man. And some have fallen under the spell of what I call Hans Rosling's 'shiny population optimism', as he almost exclusively emphasises falling birth rates without any serious consideration of what's happening in Africa and other high-fertility countries. A few still think population advocacy is wrapped up in coercive, control-based approaches.

No one can deny that there have been some repugnant, deeply authoritarian campaigns to control women's fertility (in both China and India, for example), and it's not hard to unearth racist and invariably misogynistic individuals who weave population control into their hateful world views, often with a backward look to the appalling things done in the name of eugenics. But all this is *so far* from the kind of work done today by population advocacy organisations, including Population Matters here in the UK, of which I am proud to be the president. The emphasis for NGOs operating in this space today is all on women's rights, non-coercive family planning, on the wider benefits of education for girls and of reproductive healthcare, and on ensuring that women can manage their own fertility, especially in repressive and male-dominated cultures.

Gender and climate are inextricably linked. This was powerfully articulated in an article in October 2019 in *BMJ Sexual and Reproductive Health*, which urged wide-scale increases in the use of modern contraception as part of global efforts to reduce climate emissions. It focused on the continuing horror story that 44 per cent of all pregnancies in the world today are unplanned, in large part due to insufficient availability or limited acceptance of the use of contraception. That's 99 million pregnancies that could be prevented every year, significantly improving the lot of the 270 million or so women around the world who do not have access to a choice of contraception.

None of these measures are advanced as alternatives to all the other critical decarbonisation policies we need. Of course we have to go on focusing on overconsumption, particularly in rich countries. And no one is ignoring the fact that the individual carbon footprints of people in countries with the highest average fertility are very small indeed compared to people living in high-income countries. Addressing overpopulation *and* overconsumption are mutually reinforcing aspects of a properly joined-up approach to climate change.

A remarkable new study published in February 2021 in *The Journal of Population and Sustainability*, analysed the key factors influencing total energy use and greenhouse gas emissions going back to 1990. It found that the most important driver of energy use and emissions everywhere was economic growth, but that population (which grew by two billion during that time) played a very significant part, accounting on average for a third of energy demand and carbon emissions. Improvements in efficiency which reduce both energy demand and emissions have been highly significant over the 30 years, but that population growth has meant that we're effectively 'driving with the brakes on'. The study's author comments wryly on 'the conspicuous silence in recent years about the role of population growth in the debate on environmental sustainability'.

Perhaps the most convincing case that has ever been made about the importance of family planning in the context of climate change comes through Project Drawdown ('the most comprehensive plan ever proposed to reverse global warming'), the brainchild of the indefatigable Paul Hawken. Using only peer-reviewed research, he and an inspiring team of colleagues set out to 'map, measure and model' the eighty most substantive solutions to climate change. Many are very familiar: wind and solar power, protecting forests, plant-rich diets and so on. Some have been significantly underestimated up until now – for instance, providing more sustainable and energy-efficient refrigeration (as covered in Chapter 11) comes out top of the list as the single most important solution. But for climate scientists and policy geeks, by far the most surprising solutions in the list are educating girls and family planning (captured under the heading 'Health and Education', which comes second in the list after Reducing Food Waste). Let me just spell that out: the second most effective way of driving down emissions of greenhouse gases is to invest proactively in the education of girls and ensuring women's rights, including the right of access to a choice of contraception. You don't hear that from many environmental organisations or climate campaigners. (Please check out the TED talk given by Dr Katharine Wilkinson, one of the leaders on Project Drawdown, given in November 2018, in order to get the full picture of how these things add up.)

BACK TO THE SOIL

It's to the soil that I now want to return, to end with a much more positive conclusion after what has necessarily been a grimly realistic assessment in this part of the book of some of the forces stacked against us. It's important to start taking into account the astonishing power of the Earth itself in helping us to navigate this formidably challenging transition.

Governments, agribusinesses, UN agencies and large-scale farmers have stuck with the orthodoxy of chemical-intensive farming for the past fifty years, contemptuously dismissing alternative thinking and farming practices as irrelevant to the serious business of feeding an ever-greater population. Somewhat grudgingly, the FAO used to support a global programme under the title of 'Conservation Agriculture', but that was about it by way of alternatives. As it happens, that programme turned out to be extremely successful, encouraging low-till or zero-till farming (where the soil is not regularly ploughed up after every harvest, thus reducing emissions of CO_2), using fewer fertilisers and agrochemicals, introducing cover crops to ensure the soil is never left bare, and suchlike. Farmers invariably benefited from support of this kind, reporting similar yields to what they were getting before, but with significantly reduced costs.

Right now, there's a revolution on the way, building on a lot of that early work. It's a revolution being driven by the opportunity to manage our land in such a way that billions and billions of tonnes of CO_2 can be sequestered (i.e. drawn back down) into the world's soils between now and 2050. So far, the whole climate agenda has been driven principally by the need to *decarbonise* our carbon-intensive economies; policy-makers are now (at last!) starting to plan for programmes that will systematically drive a process of *recarbonisation* – getting carbon back into our soils, wetlands, peat bogs, marine environments and so on. For me, this is one of the most exciting areas of potential transformation that is just waiting to be fully developed over the next few years.

The simplest way of describing this is 'farming carbon'. It's not a completely new idea: Australia's 'Carbon Farming Initiative', the Carbon Farmers of America and the Global Soil Carbon Challenge were all pioneers in this space. When it looked likely that the USA would be adopting a national carbon trading scheme in the early 2000s, a lot of work went into the idea of rewarding land managers for building soil carbon. Unfortunately, that all came to nothing, although it was instrumental in persuading leading US NGOs like The Nature Conservancy and the Environmental Defense Fund to start working much more constructively with farmers' organisations that might once have been perceived as the enemy. This kind of co-operation will be a critical condition of success over the next decade.

The urgency of the climate debate has now shifted things into a very different ballpark, as endorsed by a Special Report from the IPCC in August 2019 ('Climate Change and Land'), emphasising building carbon in the soil as one of the most significant climate actions available to us today. At the Paris Climate Change Conference in 2015, a new initiative was launched: '4 Per 1000' – a really obscure title based on the overall objective of increasing soil carbon by 0.4 per cent (that's four parts in a thousand) per annum. To be honest, this initiative has rather languished in the margins of subsequent climate conferences, but still has the enthusiastic backing of France and a number of other countries. A new initiative launched in 2019 in the US by Indigo Agriculture, the Terraton Initiative, has seriously raised the bar, aiming to sequester 1 trillion tonnes of CO_2 (that's what a terraton is!), and is already getting some real traction in creating a voluntary carbon market by offering to pay farmers $15 a tonne to store carbon in their soils.

All this is taking place in the wider context of what is now called 'regenerative agriculture', embracing a wide range of farming practices including the use of multi-species cover crops, crop rotations and diversification, reduced chemical inputs and zero-till. At the same time, there's a growing interest in agroforestry, which can mean anything from using trees and livestock in mutually beneficial ways, combining trees and arable crops, and planting trees to provide shelter belts or riparian buffers (to protect streams and rivers).

But how is this to be paid for? The obvious place to start is by redirecting the vast amount of taxpayers' money used to subsidise farmers today. A critical report from the OECD in June 2020 estimated that 54 leading countries are spending roughly $700 billion a year on farm subsidies – that amounts to more than 10 per cent of total farm revenues generated that year. Roughly $2 billion a day. The USA and the EU are the worst offenders.

As we saw on page 198, the vast majority of that does nothing to help protect the environment, nurture biodiversity or replenish soil quality, although the EU's agri-environment schemes have made a positive contribution on the margins of intensive farming. Farm subsidies in the USA began during the Great Depression to help farmers stay in business, and have just carried on since then – most damagingly since 1996 when the US Department of Agriculture ended a requirement that farmers should practise some kind of soil conservation. The historical perversity of this is staggering: decades of taxpayers' money being used to destroy the underlying resource (the soil) on which humankind's future prosperity entirely depends.

Beyond that, we have to look more broadly at the resources we currently commit to protecting the natural world. The United Nations Environment Programme has an annual budget of around $950 million (2019 figures) but has absolutely no executive powers over nation states. On average (according to an organisation called Government Spending Watch), the percentage of GDP devoted to environmental protection in the seventy countries it monitors has been around 0.3 per cent over the past decade, but it varies considerably between different countries. Within the EU, national expenditure on environmental protection has remained stable over the past ten years at around 1.9 per cent of GDP, with the UK at 1.8 per cent in 2017. In the USA, federal spending through the EPA (at around $8 billion a year) amounts to less than 0.3 per cent of GDP, and that fell even further under the Trump presidency. But a lot of environment expenditure takes place at state and city levels.

Beyond that, the OECD has estimated that total global annual expenditure on biodiversity amounts to a maximum of $10 billion – a fraction of the hundreds of billions of dollars that governments lay out, every year, supporting fossil fuel companies, intensive farming and industrial development.

Relatively speaking, these are tiny sums of money. Restoring the natural world, after decades of systematic over-exploitation and wanton destruction, is going to demand a great deal more of us in the future.

Part 5

ALL IN IT TOGETHER

18

TAKE HEART FROM HISTORY

*'I ain't quitting as long as I'm breathing. Hold
the ground.'*

ANGELA BLANCHARD

Climate change: as those two little words become more and
more familiar, the all-encompassing immensity of the task
ahead of us is progressively revealed, layer by layer, sector by
sector, individual life by individual life. There's never been
anything quite like this in the short but tempestuous history
of humankind – though there's still much we can learn from
earlier struggles.

First, there's an understandable tendency (especially on the
part of those who have always thought of climate change as 'an
environmental issue') to look to the record of anti-pollution
and clean-up campaigns over the past few decades to find
useful precedents. People often refer to the campaign to protect the ozone layer back in the 1980s as a recent example of
the global community successfully coming together to prevent
what could have been a massive problem for the whole of

humankind. The cause of that problem was the widespread use of a family of chlorine-based chemicals (chlorofluorocarbons, or CFCs) in refrigeration, as propellants in aerosols and so on. Once released, these gases were eating away at the ozone in the upper atmosphere, which plays a key role in protecting all life on Earth from damaging ultraviolet radiation from the sun. Once the science was established (after the usual obfuscation and prevarication from industry), the international campaign to phase out the use of CFCs (which I was very involved in as director of Friends of the Earth in the UK at that time) gained rapid momentum, with high levels of public and political engagement. The principal companies involved soon discovered that they could use other, much less damaging chemicals, and a timetable for phasing out CFCs was agreed through the Montreal Protocol in August 1987, ratified just two years later. All done and dusted over just a few years.

The latest indication from the United Nations Environment Programme is that the ozone layer won't be 'fully recovered' until after 2050 – not least because of continued flouting of the Montreal Protocol by China over the last two decades. However, the Montreal Protocol is still seen as an out-and-out success story. However, any comparison between that campaign and today's Climate Emergency simply doesn't stack up; industry was using relatively small volumes of CFCs in the early 1980s, deployed in just a few sectors. Today, we're putting more than 50 billion tonnes of CO_2 and other greenhouse gases into the atmosphere, every year, with the vast majority of people on the planet today still dependent on the use of fossil fuels.

As I've said before, an environmental framing of climate

change is entirely unhelpful anyway, and has contributed significantly to almost three decades of political indifference and neglect. Climate change is better seen as a decadal, system-wide market failure, which has comprehensively corrupted both the way we create wealth and the way we measure it. Combating that market failure is already proving very challenging, and it will get harder still, year by year, as the costs mount, measured both in billions of dollars and in human hardship. That's why more and more people now think of climate change more as a matter of human rights and social justice. In effect, a relatively tiny number of people who make their money from the extraction, processing, financing and marketing of fossil fuels, together with an equally tiny number of politicians dependent on the continuing success of those industries, are undermining the livelihoods and the life chances of the whole of the rest of humankind.

The climate crisis is utterly different from anything humankind has had to respond to before – including the COVID-19 pandemic. Not only are we all, in varying degrees, going to be on the receiving end of climate breakdown, but almost all of us, in varying degrees, are implicated in that breakdown. We all have a personal carbon footprint; we all depend on the use of fossil fuels and, more often than not, enjoy the benefits that the use of fossil fuels brings us. We are, therefore, whether we like it or not, morally bound to get involved in the necessary transformation of the global economy, a process on which the future of our species now depends. That inevitably entails a transformation of our own lifestyles, especially for those of us with high carbon footprints.

These are often painful insights for people as we gradually

come to terms with the immensity of it all. To be implicated personally, to however minute a degree, in an existential threat to the future of human civilisation (acknowledging the very real possibility of some kind of calamitous collapse, as explained in earlier chapters) is bad enough; to have to address that painful reality through the very personal lens of our own children and grandchildren (if we have them, or hope to have them) can be shocking. Unbearable even.

As I contemplate what that means for me, personally and professionally, as I think about the next stage in what has already been a long and often painful struggle, I'm drawn to a few key moments in history to see what can be learned for our work as climate campaigners – starting with the anti-slavery cause and the role of the abolitionist movement in both the UK and the USA. Wen Stephenson writes very movingly about this analogy:

> The parallels are irresistible: there's the sheer magnitude of what's at stake, in human and moral and, yes, in economic terms [...] There's the fierce opposition of powerful and entrenched reactionary forces – the Slave Power of the antebellum South and the fossil-fuel lobby of today. There's the movement's explicit emphasis on human rights and social justice, including economic and racial justice, considering that the vast majority of those suffering the earliest impacts of climate change globally are impoverished people of colour – who, let's be clear, have done nothing to create the catastrophe [...] If the abolition of slavery was the great human, moral struggle of the eighteenth and nineteenth centuries, then climate justice is the

great human, moral struggle of our own time. And the climate-justice movement has every reason to be as resolute and as radical, in its own way, as the movement that ended slavery.

AM I NOT A MAN AND A BROTHER?

These words first appeared on a medallion designed by Josiah Wedgwood, a prominent anti-slavery abolitionist, in the very earliest stages of the campaign that was launched in May 1787. A group of twelve Quakers came together in London (as the Society for Effecting the Abolition of the Slave Trade), compelled to do something about the unspeakable cruelty inflicted on millions of enslaved Africans shipped across the Atlantic Ocean to the Americas. It is reckoned that somewhere between 10 and 12 million Africans were transported between the sixteenth and the nineteenth century, causing continuing devastation in many countries in West and Central Africa. The slave trade itself was eventually abolished twenty years later, in 1807 (with the US Congress banning the import of slaves in 1808), but it wasn't until four years after the Emancipation Act in 1833 that the ownership of slaves in the West Indies *finally* came to an end. After fifty long years of campaigning. And it took longer still, with even greater misery and suffering, in the USA.

In terms of comparisons with today's climate movement, it's unnerving to look back at that first phase of activities between 1787 and 1807. Everything moved very slowly, and campaigners were careful *not* to speak out against the ownership of slaves, or to challenge the idea that it was morally acceptable for one person to own another person. More radical, morally charged voices were kept in the margins of the debate, for fear of alienating powerful establishment interests. There was huge sensitivity at that time around the sanctity of private property, and slave owners in both the West Indies and America argued very strongly that this was an economic issue, not a moral issue.

We've seen this kind of tactical divide many times in the Green movement. For instance, right from its inception in January 1980, the Green Party in Germany lived through decades of division between the 'realos' on the one hand (staying pragmatic, avoiding alienating the public, joining coalitions) and the 'fundis', with their much more absolutist views on the need for rapid and radical transformation – particularly on things like climate change. I can feel something of the same tension in debates going on within radical campaigning organisations today, with their impatience at the 'continuing failure of environmental NGOs to make any impact at all in the climate debate', deliberately ramping up more and more absolutist demands, however 'unreasonable' they might appear to be, and however ill-judged the attacks on existing organisations may be.

The tactics of the anti-slavery campaign in the UK began to harden from 1807 onwards, with the debate focused progressively on the moral case for outright abolition of slavery rather than the economic case. But the dominant tone was still very gradualist – as in the explicitly entitled 'Society for the Mitigation and Gradual Abolition of Slavery' – even as more radical voices argued passionately for *immediate* abolition. Such gradualism reflected the more conservative orientation of both Parliament and Congress; in the US, abolitionists were often looked on as dangerous, irrational radicals, and the word itself was used as an insult. Abraham Lincoln was a very cautious gradualist, and only came to accept the case for outright abolition in the run-up to the Civil War, finally influenced by the passionate advocacy of abolitionists. Just listen to the words of William Lloyd Garrison back in 1831, and reflect how one hears similar words of outrage from climate campaigners today:

> I do not wish to think, or to speak, or to write, with moderation. No! No! Tell a man whose house is on fire to give a moderate alarm. Tell the mother to gradually extricate her baby from the fire into which it has fallen. But urge me not to use moderation in a cause like the present. I am in earnest – I will not equivocate – I will not excuse – I will not retreat a single inch – and I will be heard.

Such uncompromising advocacy led directly to the UK's Emancipation Act in 1833. The debate about appropriate compensation for slave owners in the West Indies then dragged on for another four years, with the massive sum of £30 million (which is roughly £3.5 billion in today's terms) eventually being agreed. There was never any debate about compensation for the slaves. That would have been all but unthinkable at the time. Even the most fervent abolitionists in the first half of the nineteenth century had little time for the cause of racial equality; they believed that the best that could be hoped for, post-abolition, would be a slow process of 'civilising Africans' over a long period of time.

Tactical debates of this kind dominated the abolitionist

cause over fifty years, in both the UK and the US. While we know for sure that we absolutely don't have fifty years to abolish the stranglehold that today's fossil fuel super-elite has over the entire global economy, arguments about exactly what has to be done by when remain highly controversial. Things have to be done very differently in a Climate Emergency; young people involved in the school strikes are increasingly angry at the idea that the same old gradualist approaches are any longer fit for purpose.

And I fear we're going to have to face up to exactly the same sort of morally vexatious arguments over compensation for the owners of fossil fuels, as was the case with compensation for the owners of slaves. Delegations from Saudi Arabia and other oil-rich states enjoy scandalising UN conferences with their demand that, if it is somehow determined that their oil and gas must now remain in the ground, then due compensation should be paid to their governments on the basis of 'profits foregone'! One assumes they're just trying it on, but step back for a moment and imagine what is going to happen in countries like Russia, Saudi Arabia, the UAE, Iraq, Iran and Brazil as their revenues from sales of oil and gas steadily decline year after year. It's not a pretty picture.

One has to conclude any consideration of the abolition of slavery with an unavoidably grim postscript about slavery in the twenty-first century. There's no consensus on exactly how many people we're talking about here, with several different definitions. Free the Slaves keeps it simple: 'Slavery is the holding of people at a workplace through force, fraud, or coercion for purposes of sexual exploitation or forced labor so that the slaveholder can extract profit.' Based on that definition, the

Walk Free Foundation talks of 40 million people in slavery, 10 million of whom are children. Twenty million are active in the sex industry, often children. Child labour, forced migrant labour and bonded labour are still commonplace. And despite slavery now being technically illegal in every country, government-forced labour is still widespread. In China, millions of prisoners are forced to undertake unpaid labour. In the USA, the prison labour industry brings in a profit of around $1 billion a year, with prisoners often paid very little or nothing at all for their work.

In 2015, the UK was the first country to pass a Modern Slavery Act, and there's no doubt (as far as big companies are involved) that this is having a significant impact on the way they manage their supply chains. However, more than 70 years on from the anniversary of the Universal Declaration of Human Rights, it's pretty sobering to reflect on how little progress we seem to have made since the abolition of slavery nearly 200 years ago. According to Free the Slaves, a slave in the US might be sold for the equivalent of $40,000 (in today's money) just before Congress banned the import of slaves in 1808. Slaves today can be bought for as little as $100. Even now, these are long, hard battles, with many setbacks along the way. And it will be no different with the struggle for climate justice.

Women's suffrage: the fight for the right to vote

This is the second liberation struggle where there are many relevant parallels. It too took around fifty years to secure women's suffrage in the UK, and more like eighty years in the USA. And as with the campaign to end slavery, it was a slow-burn process to start with: between 1867 (when the first Societies for Women's Suffrage were established in London and Manchester) and 1897, there was plenty of activity but absolutely no progress on the political front. In 1897, the National Union of Women's Suffrage Societies (NUWSS) was set up under the impressive leadership of Millicent Garrett Fawcett. Many of the individual societies involved were pretty conservative in their orientation, resolutely middle class, and uncomfortable about doing anything that might be judged to be 'unwomanly', or to challenge prevailing gender and class norms. Their approach was very much to demonstrate the deep prejudices of those opposing women's suffrage, and to maintain the support of the general public by operating strictly within the law.

There were many, however, for whom such tactics rapidly became extremely frustrating. In 1903, the Women's Social and Political Union (WSPU) was established with an explicit goal to be as disruptive as possible in order to achieve as much publicity as possible, without worrying too much about public sympathy. Calling themselves suffragettes to distinguish themselves from the suffragists of the NUWSS, their favourite slogan was 'Deeds not Words'. From 1905 through to the start of the First World War in 1914 (when all campaigning was put on hold), any and every tactic was tried to achieve these goals – including the use of violence. Partly as a response to some very brutal and repressive tactics by the police and the government itself (especially through the force-feeding in prison of suffragettes who had gone on hunger strike), the

WSPU became more violent, condoning attacks on MPs, vandalism, sabotage and even arson. Both Emmeline and Christabel Pankhurst justified these tactics on the grounds that they had no choice since everything else had failed.

This was hugely controversial. By the time Emily Wilding Davison was killed as she ran out in front of the King's horse on Derby Day in 1913, more and more women were becoming increasingly vocal in their rejection of such tactics. These disputes were only resolved because of the First World War. As soon as it ended, a partial victory was achieved in the 1918 Representation of the People Act, when all women over the age of thirty won the right to vote, with seventeen women standing as candidates in that year's general election. Equal representation was finally secured in 1928.

There are many parallels between this extraordinary period in British history and the way in which today's climate campaign is evolving. It's been argued that the period of time between the United Nations Framework Convention on Climate Change being signed off at the Earth Summit in 1992 and the global climate strike on 20 September 2019, when around 100,000 people took to the streets of London, is pretty similar to the first thirty years of the women's suffrage movement between 1867 and 1897: growing public support, impressive civil society organisation (including lots of 'establishment backers'), an increasingly authoritative intellectual rationale – but relatively little to show for all that effort, with the one outstanding exception of the passing of the UK's ground-breaking Climate Change Act in 2008.

Extinction Rebellion (XR) explicitly references the absurdity of sticking with the same tactics (what it calls 'the long march through the institutions') when those tactics are seen by them to have achieved so little. Many don't agree with that judgement – including myself. But I *do* agree that more radical campaigning, specifically designed to change the rules of the game, to persuade people to stop pretending that leisurely incrementalism is any longer an appropriate response as the world gets warmer and the weather gets more violent in front of our eyes, is absolutely necessary.

There is of course one difference – a very big difference – between XR and the WSPU. So great was the anger and frustration of the WSPU that its members condoned and even encouraged the use of violent tactics. XR is, as we saw in Chapter 9, resolute in its total rejection of the use of violence. Though any organisation with such loosely federated governance arrangements as XR's is always vulnerable to entryism from those who may have no such hesitation about using violence, XR's resolve in this area remains solid. Without it, many of those 'ordinary citizens' who felt compelled in 2019 to bear witness by being prepared to get arrested would be far less likely to come forward.

As we saw with the controversy in October 2019, as XR protestors disrupted commuters using public transport to get to work (see page 127), the risk of backlash is never far away. This was a huge issue for suffrage campaigners in the UK from 1889 onwards, with many MPs either directly or indirectly supporting the National League for Opposing Woman Suffrage. In 1908, the Women's National Anti-Suffrage League was established, and succeeded in raising 400,000 signatures calling for an end to the suffrage movement. As it happens, XR's resolute support for non-violent direct action more closely mirrors the tactics of the Women's Freedom League than those of the WSPU. Following the example of Mahatma Gandhi, the League urged its supporters to refuse to co-operate, to be prepared to break the law and to be arrested, but never to resort to violence.

Much to learn, that's for sure. Alice Paul, one of the most constant champions of women's suffrage in the USA, famously said: 'When you put your hand to the plow, you can't put it down until you get to the end of the row.' But it's a very different challenge confronting today's climate campaigners. There's no 'end of the row', no one point somewhere out there in the future where victory can be declared, however wearily. There's no single Act of Parliament or passage of a bill through Congress that will mark *the* definitive turning point. The juggernaut that is today's 'industrial growth economy', powered predominantly by fossil fuels, animated by an inhumane and soulless consumerism, can indeed be slowed and then turned — the abundance of solutions provides ample proof to that effect. That's what I've tried to emphasise all the way through *Hope in Hell*. But we're still in the very early days of what will inevitably be a very, very long-drawn-out campaign.

19

A JUST TRANSITION

'Protest that endures is moved by a hope far more
modest than that of public success, namely, the hope
of preserving qualities in one's heart and spirit that
would be destroyed by acquiescence.'

WENDELL BERRY

The principal difference between today's Climate Emergency and world-changing, historical moments like the abolition of slavery or votes for women is 'the universality' of climate change (there will be not one single human being whose life will be untouched by the consequences of the Climate Emergency), and its sheer, all-but-incomprehensible scale. Writing in the *Los Angeles Review of Books* in June 2019, author and army veteran Roy Scranton put it like this:

Climate change is bigger than the New Deal, bigger than the Marshall Plan, bigger than World War II, bigger than racism, sexism, inequality, slavery, the Holocaust, the end of nature, the Sixth Extinction, famine, war, and plague all put together, because the chaos it's bringing is going to supercharge every other problem. Successfully meeting this crisis would require an abrupt, traumatic revolution in global human society; failing to meet it will be even worse.

As with every other aspect of today's global economy, this is also a story of grotesque inequality. Oxfam regularly reminds us that the richest 10 per cent of people in the world today are responsible for at least 50 per cent of consumption-related greenhouse gas emissions; the poorest 50 per cent of people today account for just 10 per cent of consumption-related emissions. Beyond that, we already have ample evidence that the Climate Emergency acts as 'an amplifier' of social inequality, disproportionately affecting the poorest and most vulnerable in society. It's simply impossible to imagine humankind successfully navigating the decades of radical decarbonisation ahead of us without addressing the deep, structural inequalities that blight the lives of so many billions of people today. This reality is increasingly captured in the idea of a Just Transition.

This reflects some of the painful learning of the past decade or more that no progressive social movement can effectively advance its cause without explicitly addressing the concerns of 'the left-behind' – including those who may very well become left behind in the future through an accelerated transition out of fossil fuels. It's highly significant that this factor features prominently in both the proposed Green New Deal in the USA (which has committed unequivocally to protecting the interests of those in the fossil fuel industries who lose their jobs) and in the EU's €17.5 billion Just Transition Fund to support those European nations most vulnerable to the impacts of rapid decarbonisation, particularly Poland (which employs more than half of the 230,000 people working in Europe's coal sector), Germany and Romania, with their continuing dependence on coal. The EU Commission's new president, Ursula von der Leyen, has spelled this out: 'We want to be front-runners in climate-friendly industries, in clean technologies, in green financing, but we also have to ensure no one is left behind. This transition will either be working for all, or it will not work at all.'

The coal sector alone employs 9 million people globally. In South Africa, some 200,000 miners play a crucial role in the economy, but 10,000 have already lost their jobs and many more are vulnerable as South African energy companies shift to renewables. The social and economic consequences of this transition being poorly managed (or not managed at all) could be politically explosive. Even Germany, which set up a Commission on Growth, Structural Change and Employment in 2018, is struggling to come up with recommendations and timelines that its many different stakeholders can sign up to.

There's an extra dimension to this in the USA, where a well-established 'environmental justice' movement has exposed some of the limitations of the mainstream, Washington-focused environmental NGOs. Back in 1994, Robert D. Bullard founded the Environmental Justice Resource Centre at Clark Atlanta University, arising out of nearly two decades' work revealing the 'structural environmental racism' that led to the siting of petrochemical plants, refineries, landfills, waste incinerators and power plants close to communities of colour, especially African American communities in the South. People of colour are more likely to live in areas that are most vulnerable to flooding, and less likely to have the funds to recover from extreme weather events. Mainstream NGOs in the North, predominantly middle class, white and 'establishment-facing', never got on this particular page, causing some to question how well-placed they are today with 'climate change looming as *the* environmental justice issue of the twenty-first century' – in Robert Bullard's words. Their collective world view sometimes seems very far removed from the raw, cumulative anger of those disadvantaged communities who've been on the front line of environmental campaigning for the past five decades.

COVID-19 exposed these failings even more starkly, and environmental factors played a big part in that. In many US cities, asthma rates in predominantly Black communities are more than double those in largely White areas. Black Americans are more than three times as likely to die from pollution-related causes. COVID-19 has been both an environmental justice and a public health crisis. Twice as many Black Americans have died in the pandemic than White Americans.

This was explicitly recognised in the Biden/Harris Climate Plan. Vice-President Harris started an Environmental Justice Unit when she was District Attorney in San Francisco 15 years ago, and introduced a Climate Equity Act (together with Alexandria Ocasio-Cortez) in August 2020. The Climate Plan was particularly strong on environmental justice, promising a much clearer focus on 'those threatened by the cumulative stresses of climate change, economic distress, racial inequality, and multi-source environmental pollution.' 40 per cent of planned investments in clean energy, affordable housing, energy efficiency schemes, and so on, will be earmarked for use in disadvantaged communities.

FAIRNESS IN A POST-GROWTH WORLD

It's clear to more and more people across the developed world, regardless of their party affiliation, that the current version of neoliberal capitalism is starting to fail even in its own terms. Rather than a widely shared prosperity, enabled by the agile, efficient workings of a properly regulated market, it has produced chronic wage stagnation, with increasing numbers of people in work sliding deeper into poverty, with widening disparities in income, opportunity and social mobility, instability in capital markets, the rise of populism, and, now, the twin crises of the Climate Emergency and ecosystem collapse. Many senior rightwing politicians in both the US and the UK have started to express real concern about the implications of this for social and economic cohesion – and indeed for the future of capitalism itself.

On both sides of the Atlantic, there are now many commentators thinking through what lies on the other side of this particularly damaging neoliberal version of capitalism, a debate now amplified by the traumatic economic impact of the COVID-19 pandemic.

In September 2019, the Institute for Public Policy Research recommended that the government should start taxing income from *wealth* (shares, property, fine art and other investments) at exactly the same level as it taxes income from *work*. At the moment, capital gains tax in the UK is anywhere between 20 and 28 per cent. As regards income tax, higher-rate taxpayers pay 40 per cent on amounts above the basic tax rate, rising to 45 per cent on annual income above £150,000. Tom Kibasi, director of the IPPR, suggested that levelling up these two different kinds of tax should be accepted by everyone as 'a matter of basic fairness'.

This is a pretty modest proposal, and there are many other distortions in the tax system that will need to be addressed to ensure any kind of genuine fairness. This is going to be all the more important in an economy that is not growing at the rate that people have come to expect. As Professor Tim Jackson (whose *Prosperity Without Growth* – revised in 2017 – remains one of the most important texts ever written in this area) continues to point out, the treasuries and finance ministers of all OECD countries have literally no competency in thinking through the practicalities of a (permanently) low-growth economy. All their macroeconomic modelling still rests on assumptions of high economic rates of growth extending indefinitely into the future, at a time when it's increasingly clear that such growth cannot possibly be sustained ecologically.

Despite years of remorseless indoctrination, there is absolutely no reason to see permanent economic growth (which is physically impossible anyway, on a finite planet) as being the only way to ensure progress in improving people's wellbeing. In fact, there's more and more evidence to show that these two things are not only *not* dependent on each other, but that the current kind of growth is seriously impacting more and more people's wellbeing. For instance, even though the US economy is five and a half times bigger now than it was in

1969 (in terms of real GDP), American citizens are actually losing ground on the annual Happiness Index.

People's readiness to embrace post-growth ideas (which will inevitably sound pretty scary given our fixation on the notion of progress delivered uniquely through economic growth) will depend in large measure on how secure they feel in their own lives and communities, and in terms of fair access to the fruits of the economy. Right now, it's widely recognised that more people feel more insecure about their own economic prospects than ever before in both the UK and the US, and that is going to have to change. In 2019, the New Economics Foundation published a fascinating book by Anna Coote and Andrew Percy, *The Case for Universal Basic Services*. The basic premise couldn't be more simple: that everyone is entitled to services that are sufficient to meet their needs, regardless of their ability to pay, including not just existing services such as healthcare and education, but social care, housing, transport and access to the digital economy.

The New Economics Foundation's estimate for the cost of providing universal basic services in OECD countries is between 4 and 5 per cent of GDP, which would rise to around 6 per cent if today's guaranteed income protection schemes (social security, child benefit, etc.) are retained – and restored to 2010 levels in real terms. That's a high price to pay for any OECD country, but it speaks very powerfully to this notion of 'healing' that is now so important in such deeply polarised countries as the UK and the US. For me, that leads to one clear conclusion: it's impossible to provide the kind of public services we so urgently need without people being prepared to pay higher taxes.

SOCIAL COHESION IN A FRAGMENTING WORLD

Even without a Climate Emergency to deal with, the question of how best 'to maintain our humanity' looms ever larger in the USA, the UK and almost every country where hugely divisive forms of populism have stepped into the gaps left by the manifest failures of neoliberal capitalism. My good friend David Fleming was prescient in recognising the challenge this would pose to our societies as the old order met fewer and fewer of people's expectations, but with nothing to take its place. His wonderfully unorthodox proposition was to trust far more to people making it all work in their own communities, to powerful historic drivers of self-help and collective solidarity, working in mutually reinforcing ways (through local support groups, credit unions, volunteering schemes and so on), to conviviality, culture and common civility speaking to the best in each of us rather than leaving people increasingly vulnerable to today's stripped-down, devil-take-the-hindmost individualism. He constantly reminded us, right up until his untimely death in 2010, that if conventional economists were ever sad enough to monetise the combined 'value' of the millions of person-days committed to voluntary activities at every level in society, including the selfless devotion of tens of thousands of carers looking after parents, grandparents or children, it would give an infinitely bigger boost to GDP than any 'stimulus package' the Treasury could ever come up with.

But David worried away at how we might best protect all of that in the coming crisis/breakdown, which he saw as inevitable. I'm worried about that too.

One of the liveliest debates among environmentalists back in the 1970s (when David and I were both involved in the Green Party) revolved around the potential incompatibility between democracy and the imperative of living within our ecological means. Writers like Garrett Hardin, William Ophuls and Robert Heilbroner were much taken by a quote from Edmund Burke written in 1791: 'Men are qualified for civil liberty in exact proportion to their disposition to put moral chains upon their own appetites [...] Society cannot exist unless a controlling power upon will and appetite be placed somewhere; and the less of it there is within, the more there must be without.' Today, as the 'fairy tales of eternal economic growth', in Greta Thunberg's words, start to vanish in the mist, there is a very real fear that current levels of inequality will drive even greater resentment, even deeper anger. And it's clear that one of the flashpoints for this will be immigration.

This is certainly not the place to address in any detail such a complex and controversial area of concern. Back in 2017, myself and Colin Hines (a veteran environmental campaigner) co-authored an essay, 'Getting Real about Immigration', inviting people who position themselves as 'progressives' on the political spectrum to accept the overarching need for the EU (and now for the UK outside the EU) to find more effective ways of managing migration into the EU that run counter to the prevailing views of most progressive voters. On the one hand, we must emphatically reject the cruel and often 'institutionally racist' positions taken by the UK's Home Office; but on the other hand, we still have an obligation to join up the dots here in terms of demographics and the Climate Emergency.

We know, for instance, as a matter of increasingly painful

inevitability, that the lives of tens of millions of people (and possibly hundreds of millions of people) will be devastated by the effects of climate change. Particularly in Africa and the Middle East. Unless we start doing things very differently, we know that many of those people will have no choice but to leave their homes and communities if they are to have any prospect of survival, let alone a better life. And we know that many of them will seek to come to Europe, as the place that offers the best possible refuge in an all-encompassing storm that is not of their own making.

We have to go back to those uncomfortable population projections on page 265. Contemplating the possible consequences of that kind of growth in Africa, against the backdrop of our rapidly worsening climate and biodiversity emergencies, Jeremy Grantham, one of the most far-sighted and progressive private asset managers anywhere in the world, had this to say in 2018:

> I wrote five years ago that the first casualty of this African problem would be the liberal traditions of Europe. Well, it happened a whole lot faster than I feared! Just an accumulated couple of million refugees are already providing political propaganda that is empowering right-wing groups everywhere in Europe. Imagine if Europe were to try and take 100 million who will want to emigrate if the population keeps growing like this. Europe will need to get its act together, and form a joint policy that is as gentle and as firm and as reasonable as it can possibly be. It simply will not be able to take and absorb nearly as many food and climate refugees as would be required to solve the problem.

STRATEGIC ADAPTATION

It's uncomfortable hearing projections of that kind. Boiling it down to gross numbers, rather than responding more empathetically to the suffering of the individuals behind those numbers, can be very alienating. But we cannot shy away from a 'future reality', which the science of climate change shows us is become more and more likely, just because it makes us feel so uncomfortable.

We do indeed have to get our heads around this if we Europeans are to play our part in ensuring a genuinely just transition. And the same is true in the United States, contemplating the terrible damage that today's Climate Emergency will inflict on the people of Central and South America. In September 2019, the International Federation of Red Cross and Red Crescent Societies issued a devastating report ('The Cost of Doing Nothing') showing that 2 million people *a week* already need humanitarian aid because of the Climate Emergency, and that this number will double by 2050 (to around 200 million every year) if governments fail to act now. Its principal recommendation was to increase the resilience of vulnerable people through improved early-warning systems, restoring natural features such as mangroves, swamps and wetlands, off-grid renewable energy, drought-resistant crops, rethinking housing and infrastructure, providing shelters and so on – in other words, through timely, precautionary adaptation.

This emphasis on strategic adaptation (as promoted by the Global Commission on Adaptation, launched in October 2018, with its first report published a year later) will be a particularly crucial element in any just transition, with significant implications for all nation states. All foreign policy, all trade agreements and all aid and development transfers will need to be focused on protecting vulnerable people in the places where they live. Aid and development policies must prioritise investments in human and social capital (particularly education for girls, access to contraception and reproductive healthcare) and in natural capital – soil, water and biodiversity. And if we're serious about both our own national security, and that of all those we're working with to promote strategic adaptation of this kind, arms sales will need to be dramatically curtailed, and a percentage of defence spending reallocated to help countries protect those all-important natural defences.

Talking of natural defences, there's one aspect of a just transition that needs a great deal more attention from scientists, policymakers and Western NGOs: the pressing need to protect indigenous people around the world. In many countries, it's indigenous people who are on the front line of protecting forests and other ecosystems, and they pay a very heavy price for this. Figures from organisations such as Global Witness and Front Line Defenders show that at least 212 environmental rights defenders were killed in 2019, up from 164 in 2018, many of whom were indigenous people protecting their land from the impacts of mining and intensive agriculture. The true number is likely to be much higher as many such offences go undocumented, and perpetrators almost always go unpunished. Colombia, the Philippines and Brazil are the countries with the highest death rates. There are fears that 2020 will prove to be an even worse year, with many governments using the pandemic to crack down even more harshly on campaigners' freedoms.

The Environmental Justice Atlas reckons there are now more than 3,000 conflicts over land and resources worldwide, many of which involve notionally 'protected indigenous territories'. More than 28 per cent of the global land area is owned, used or managed by indigenous peoples, including more than 40 per cent of officially protected terrestrial areas. These areas are home to an astonishing 80 per cent of the Earth's biodiversity.

One of the more unfortunate consequences of the pandemic in 2020 has been its impact on some of the poorest but most biodiverse regions in the world, with significant economic disruption caused by the loss of revenue from ecotourism and conservation initiatives. It's reckoned that ecotourism provides more than 80 per cent of the income for National Park agencies.

Without the same high level of protection, there's been a marked increase in poaching in areas where there would normally be too many tourists for poachers to operate safely. As New Scientist's reporter Donna Lu has pointed out: 'It's a timely reminder of how the most effective solutions to the biodiversity crisis are human ones. Protecting Earth's precious ecosystems means empowering the people who are closest to them.'

Regrettably, one additional source of pressure on indigenous people has been the conservation movement itself. Protected areas and national parks have often been created by taking land away from indigenous people, either evicting them completely or excluding them from conservation programmes.

They were routinely seen as 'obstacles to conservation', and in a pattern of behaviour over many years that can only be described as explicitly racist, their deep knowledge was completely ignored. Even now, WWF International is embroiled in a number of controversies (exposed in 2019 by BuzzFeed News) involving attacks on indigenous people around parks and other conservation areas in both Asia and Africa.

The evidence regarding the success of indigenous people in protecting their land tells us the real story. Research shows conclusively that nature is declining less rapidly on land managed by indigenous peoples – particularly in countries like Brazil, Colombia and Bolivia. In something of a breakthrough last year, the critically important contribution of indigenous people to conservation and 'alternative science' was explicitly cited by the highly influential report from the Science-Policy Platform on Biodiversity and Ecosystem Services, which I referred to on page 259. This includes some explicit recommendations on tackling the marginalisation of indigenous communities through: '[...] national recognition of land tenure, access and resource rights in accordance with national legislation, the application of free, prior and informed consent, and improved collaboration, fair and equitable sharing of benefits arising from the use and co-management arrangements with local communities'. This powerful endorsement will be enormously helpful to the likes of Survival International, which has battled away for more than fifty years defending the rights of indigenous people all around the world – with mighty little support from environmental organisations during that time. The tide is now definitely turning.

This explains in part why there is now growing interest in

the idea of establishing the crime of 'ecocide' under the Rome Statute of the International Criminal Court. The Rome Statute was signed by 123 nations in 1998 seeking to prosecute 'crimes against peace': genocide; crimes against humanity; war crimes; and crimes of aggression. At the time, it was proposed that there should be a fifth 'crime against peace' – ecocide. This is defined as: 'the extensive loss or damage or destruction of ecosystem(s) of a given territory, whether by human agency or by other causes, to such an extent that peaceful enjoyment by the inhabitants of that territory has been or will be severely diminished'. In essence, this would create a legal duty of care for life on Earth, and would radically shift the balance of power in today's suicidally destructive global economy. For reasons that were not clear at the time, proposals for that fifth crime against peace were dropped at the last moment.

Because of the way the Rome Statute is organised, any signatory nation can propose an amendment which cannot be vetoed. Member states can only sign or abstain, and when two-thirds of member states have signed it, it becomes law. And that's exactly what the organisation called Stop Ecocide is intent on making happen. In December 2019, a very significant 'first step' was achieved. At a meeting in The Hague of the 'States Parties' to the Rome Statute, both the government of Vanuatu in the Pacific and the government of the Maldives in the Indian Ocean moved an official amendment to the Rome Statute that would criminalise acts that amount to ecocide. In February 2021, the European Parliament voted to urge 'the EU and its Member States to promote the recognition of ecocide as an international crime under the Rome

Statute of the International Criminal Court', a decision reinforced by the Parliament's Environment Committee asking that the Commission 'should study its relevance for EU law and diplomacy'.

Just a start, and there's still a very long way to go. But as the idea of a just transition gains ground, featuring more and more prominently in international processes and in front-line campaigns, it seems only right that formal judicial process should be part of the mix, at the highest possible level. As Stop Ecocide says: 'Serious harm to the Earth *is* preventable. When government ministers can no longer issue permits for it, when insurers can no longer underwrite it, when investors can no longer back it, when CEOs will be held criminally responsible for it, the harm will stop.'

That's precisely the kind of message of hope we all need to hear in these extremely troubling times. Particularly young people.

20

MOMENTS OF TRUTH

'We are caught in an inescapable network of
mutuality, tied in a single garment of destiny.
What affects one directly, affects all indirectly.'

MARTIN LUTHER KING

Writing this book has caused me more pain than any other I've written. *The World We Made* had plenty of hard-hitting things to say about climate change and ecosystem collapse, but by virtue of 'cheating' on the chronology (writing it from the perspective of someone in 2050 looking back on all the amazing things that made it possible for people to be enjoying a good life at that time), I was also able to cheat on my own emotions, keeping at bay more apocalyptic accounts of why such a world may no longer be available to us in 2050, to avoid confronting, in Roy Scranton's memorable phrase, 'The all-too-real possibility that the story we're living is a tragedy that ends in disaster, no matter what.' With *Hope in Hell*, I've had no choice but to confront that possibility, head-on, eschewing any convenient psychological refuges, testing almost to destruction my own hypothesis that there *is* still hope. That there *is* still time. That we already have *all* the solutions we need.

Susie Orbach, one of the UK's leading psychotherapists, writes about this very movingly in *This Is Not a Drill*. Her words are shown here in a poetic form by Anthony Wilson:

To come
into knowing
is to come
into sorrow.

A sorrow
that arrives
as a thud,
deadening

and fearful.
Sorrow
is hard
to bear.

With sorrow
comes grief
and loss.
Not easy feelings.

Nor is guilt,
nor fury,
nor despair.
Climate sorrow

Opens up
into wretched
states of mind
and heart.

We can find it
unbearable.
It is hard,
very hard,
to stay with.

Many people still avoid thinking too much about climate change, not because they're 'in denial' or don't care, but because it can indeed be overwhelming. Some have described this phenomenon as 'disavowal', keeping difficult stuff at bay so that it doesn't interfere too much with people's everyday life. Psychotherapists and mental health workers are reporting more people expressing 'eco-anxiety', and there's been a proliferation of articles and self-help manuals providing homely advice on ways of avoiding eco-anxiety by focusing as much on all the good things going on as on all the bad things, even to the extent of consciously managing one's social media and news feeds to maintain some kind of psychological balance.

I don't think there are any short cuts here, especially when it comes to plumbing the depths of what has been called 'anticipatory grief', empathising with the suffering and trauma still to come as a consequence of the now inevitable disruption of the lives of hundreds of millions of people. Few of us are even minimally prepared, emotionally or psychologically, for what is coming. By contrast, for those who have come to a settled conclusion that it really is too late to avoid runaway climate change, there is both the deepest pain but also complete clarity. One of the most remarkable essays I've read over the past few months is 'Facing Extinction' by Australian campaigner Catherine Ingram, who writes about 'experiencing anticipatory grief for everyone':

> We grieve because we love. To the degree that your heart is shattered over loss is precisely the degree to which you loved that which has gone. We know that coming to terms with one's own personal death or the death of a loved one can lead to acceptance. But no matter how clear and rational our understanding of the situation, many of my extinction-aware friends admit that the magnitude of the loss we are undergoing is unacceptable to the innermost psyche. It might be akin to a parent losing a young child. Even when there was no one to blame, and no story of 'if only', the sorrow can rarely be overcome. Only this time it is all the little children. All the animals. All the plants. All the ice.

Even though I do not share her 'categorically too late' position, I greatly admire her determination to focus on what really matters, 'to make life still relevant and even beautiful', with six punchy bits of advice: release dark visions of the future; find your community (or create one); find your calm; be of service; be grateful – 'longevity was never a guarantee for anyone at any time in history'; and give up the fight with evolution. It's only this last maxim that I want to push back against as hard as I possibly can. Here's what she says: 'Give up the fight with evolution. It wins. The story about a human misstep in history, the imaginary point at which we could have taken a different route, is a pointless mental exercise. Our evolution is based on quintillions of Earth motions, incremental biological adaptations, survival necessities and human desires. We are right where we were headed all along.'

For me, that takes climate fatalism to new heights. I've no doubt that campaigners for the abolition of slavery, for women's votes and for universal suffrage, for civil rights in the USA, for the end to apartheid, for the demolition of the Iron Curtain at the end of the Cold War, for the rights of LGBT people, and for literally thousands of less celebrated campaigns that have so markedly improved the lives not just of those directly affected but for all of us over subsequent generations, would have had to put up with more than their fair share of similar 'get over it' head-patting. I want to address just how misguided I think such a position is in the next chapter.

But first, a word about 'human desires'. The lure of life-destroying economic growth is still pitched at people the world over (in the rich world and the poor world alike) as the best way of fulfilling people's desires and increasing their overall quality of life. But it's no longer delivering. In August 2018, *Observer* journalist Will Hutton wrote an article about a new syndrome being referred to by doctors in both the US and the UK to describe 'a tangled mix of economic, social and emotional problems, which consists of low mood caused by adverse life circumstances [. . .] "Shitlife syndrome" (SLS), as the doctors call it, is when finding meaning in life is close to impossible, and the struggle to survive commands all intellectual and emotional resources.'

That provides a pretty grim backdrop to the lives of so many people, particularly young people. These problems are primarily the symptoms of much deeper concerns about economic insecurity, changes in the labour market, poor-quality housing and so on. I really don't want to come across all Pollyannaish about this, but might there not be a massive sense of collective relief and liberation in coming to the conclusion that we can no longer hitch our wagon to the idea of seeking better lives through more and more consumerism? We've been brainwashed into thinking that consumerism is pretty much 'the peak of civilisation', and that there really are no alternatives other than looking to more of the same. As Rupert Read says: 'As these messages are endlessly repeated and normalised, our imaginations begin to contract, and we lose the ability to envision different worlds, different ways of living and being.' I'm fascinated by how many people are now commenting on the way in which this consumerist culture can be held directly responsible for 'an apathetic sadness of the soul', with daily reminders that consumption driven by Fear Of Missing Out, by Instagrammable likeability, by competitive excess (as in 'my stag/hen party and wedding is bigger and so much more expensive than yours'), or by monetised proofs of love, cannot possibly, *ever*, satisfy the deep craving that all of us have for meaning and purpose.

This is nothing short of pathological, in that it's more and more apparent just how much harm it's doing to us today, let alone to all future generations. Addictive consumerism, imposing an ever greater burden on the Earth's resources and ecosystems, cannot possibly provide a stable foundation for the future of humankind. We are not unique in that particular pathology; Jared Diamond's *Collapse* (published in 2004) catalogues a portfolio of historical instances of societal collapse, particularly those involving significant environmental factors. People tend to forget the subtitle of *Collapse – How Societies Choose to Fail or Succeed*, pertinently reminding us that we *still* have a choice should we choose to exercise it. Many of those who come to the conclusion that it is indeed too late to avoid runaway climate change doubt that we're capable of exercising any such choice. This is how Roy Scranton reflects on that civilisational reality:

> Civilisations have throughout history marched blindly towards disaster, because humans are wired to believe that tomorrow will be much like today – it is unnatural for us to think that this way of life, this present moment, this order of things, is not stable and permanent. Across the world today, our actions testify to our belief that we can go on like this for ever, burning oil, poisoning the seas, killing off other species, pumping carbon into the air, ignoring the ominous silence of our coal mine canaries in favor of the unending robotic tweets of our new digital imaginarium.

As I suggested in the Introduction, today's civilisational reality, COVID-shrouded as it is, looks and feels very different to the way things were in March 2020. The pandemic has fired a powerful warning shot across our bows in a manner that nothing else could possibly have done. The previously unimaginable (in terms of the lives of every human being on the planet being traumatically disrupted) became not just imaginable but a lived reality. We've experienced at first hand

a global shock caused by our disregard for the natural w
exacerbated by the complex interdependencies of the glob
economy, and demanding of us an unprecedented level of
global cooperation – a recognition that no-one is safe until
we are all safe.

For those of us in global north, we've also experienced a sense of vulnerability and fragility as never before.
Throughout the pandemic we've found ourselves fearing the
worst, for ourselves and for our loved ones. 'Intimations of
mortality' means something different now than it did before.
Erstwhile certainties about jobs, holidays, investments, and so
on, have become contingent possibilities.

Premature though this may sound, this new-found awareness is the gift of COVID-19. Just so long as we're able to
hold on to all that experience, painful though much of it is,
in the front our minds rather than pushing it away from us
as fast as we possibly can. Just so long as we're able to apply
all that learning to addressing what are potentially even more
traumatic global shocks just around the corner. Just so long as
humility takes the place of hubris, and fruitful cohabitation
with nature takes the place of increasingly dangerous 'command and control' fantasies.

The future of humankind may just depend on us embracing
the gift of COVID-19 for what it really is.

FAITH IN NATURE

And I can't help but believe that religious and faith leaders
may do a better job in that regard than political and business leaders.

Interestingly, one of the most outspoken critics of today's reckless, turbocharged capitalism is Pope Francis. He entirely rejects the idea of an economy based on the ceaseless satisfaction of individual demands, where greed and affluence are condoned so long as they help grease the wheels of a completely amoral market. In his powerful encyclical in 2015 (*Laudato si': On Care for Our Common Home*) he wrote as follows:

> When people become self-centred and self-enclosed, their greed increases. The emptier a person's heart is, the more he or she needs things to buy, own and consume. It becomes almost impossible to accept the limits imposed by reality. We should be particularly indignant at the enormous inequalities in our midst, failing to see that some are mired in desperate and degrading poverty, with no way out, while others have not the faintest idea of what to do with their possessions, vainly showing off their supposed superiority.

Six years on, *Laudato si'* is still an extraordinary document, articulating a hugely powerful case for accelerated action on climate change, challenging centuries of Catholic teaching about the supremacy of the human race, speaking of the natural world and all its creatures with genuine love and passion (with language shaped as much by the indigenous cosmologies of Latin America, centred on a living and sacred Earth, as on the usual 'domination relationships' so familiar to all Christians), with an entire chapter on the critical importance of 'an ecological conversion' among Christians: 'Living our vocation to be protectors of God's handiwork is essential to a life of virtue; it is not an optional or a secondary aspect of our Christian experience.' More recently, in November 2019, speaking at the International Association of Penal Law, Pope Francis indicated that he was thinking of adding 'ecological sin against our common home' to the Catholic catechism; it would be hard for Catholics not to be influenced by acts of environmental desecration being officially categorised as sinful. As he says, 'Creation is groaning'.

There's much about the Catholic Church that remains deeply problematic. Its historical opposition to the use of contraception is quite simply absurd – and still hugely damaging to the lives of millions of women. Its continuing failure to hold to account its own senior clerics for the systematic abuse of children over many decades is morally repugnant. And there are many Catholic bishops (particularly in the USA) who have been quick to dismiss *Laudato si'* and Pope Francis's leadership on climate change. But that leadership will, I believe, become more and more important, as 1.2 billion Catholics start to examine their own conscience more rigorously as the ravages of climate change impact people's homes and communities around the world.

By comparison, the approach of the Church of England can only be described as anaemic. Despite lots of eloquent homilies, alongside well-meaning and often impressive action taken by committed churches, with many individual Christians very actively involved in local campaigns, the Church of England itself still seems to be hanging back, and is still failing one critical test: divesting all of its £8 billion investment fund from fossil fuels. At the time of writing, the assets committee of the Church Commissioners still favours a softly-softly approach, getting out of investments in coal and the tar sands, but continuing to engage with mainstream fossil fuel companies before divesting in order 'to test whether or not they're taking their responsibilities seriously'. Oil companies couldn't care less about such representations, and it's sad to see the Church of England mired in such sophistry and tawdry compromise.

With 1.8 billion adherents, Islam is the world's second largest religion after Christianity, but has yet to become a major 'force for good' in terms of addressing today's Climate Emergency. However, they undoubtedly have the 'theological wherewithal' to take on that challenge, with many interfaith scholars pointing out that environmental thought is central to Islamic teaching, with around 750 verses of the Qur'an (roughly one-eighth of its total length) reflecting on natural phenomena and a wider environmental sensibility. In 2015, an very eminent group of Islamic scholars produced the 'Islamic Declaration on Global Climate Change', a powerful and comprehensive document based on the notion of '*mīzān*', or equilibrium in God's creation:

We affirm that:

- God created the Earth in perfect equilibrium (*mīzān*);
- By His immense mercy we have been given fertile land, fresh air, clean water and all the good things on Earth that make our lives here viable and delightful;
- The Earth functions in natural seasonal rhythms and cycles: a climate in which living beings – including humans – thrive;
- The present climate change catastrophe is a result of the human disruption of this balance.

There's no doubt that such an approach can play an important role in a country like Indonesia, where Muslims make up around 87 per cent of that country's 276 million people. The Indonesian Council of Islamic Scholars has developed a number of important religious edicts on environmental conservation, and there's definitely been an increase in the number of 'Green Mosques', as well as religious groups combining forces with local conservationists to protect endangered areas of special importance, blending 'science and numbers' with an appeal to religious values and teaching. But there's not much evidence, as yet, that this has had any discernible impact on preventing deforestation or addressing other big environmental problems.

Like it or not, our world is still a religious place, and one can point to equally significant developments in all major religions. The quality of leadership in this sphere of people's lives is likely to become more and more important as accelerating climate change bites deeper into our lives. As writer and BBC producer Mary Colwell says:

> Faiths interpret the world through story, poetry and art, requiring an imaginative approach to the Earth that works at the level of the heart. Humanity has communicated great truths in this way long before science. Faiths also use ideas that tend to be missing from purely scientific approaches, such as joy, kindness, self-sacrifice, simplicity, hope, obedience, awe and wonder – all of which are universally felt and understood.

2021 will be a critical year for faith groups to make their voices heard at both the Kunming Conference on Biodiversity in

October, and the Climate Change Conference in November. In March, the Green Faith International Network organised a day of action with more than 400 events in 50 countries, supported by 230 high-ranking religious leaders and 130 religious groups of every denomination, representing around 100 million people. That's not a bad start!

SPIRITUAL TRUTHS

Beyond religion, there are many people who profess a deep sense of spirituality (that 'awe and wonder') in their relationship with the natural world. The climate crisis goes right to the heart of who we are, and demands of us a response that cannot be entirely met by confronting the science, or committing to political and campaigning action. In many ways, the climate crisis is a spiritual crisis, laying bare the cumulative consequences of allowing ourselves to have become so completely disconnected from each other, from the world around us, and from our basic responsibility for all those who come after us. In interrogating that unfortunate legacy, we cannot avoid talking about our most fundamental principles and values, about our vision for the world.

Here again, we have much to learn from indigenous people around the world. One of the most inspiring manifestations of the fight against the fossil fuel industry in the USA has been the role of Native Americans in protesting against pipelines and other new developments. Sacredness of the land has been the unifying principle behind these protests, particularly the Standing Rock protest in September 2016 seeking to stop construction of the Dakota Access Pipeline, stretching over

1,170 miles through many areas of land held to be sacred by Native American groups.

Exactly the same spirit can be seen in the concept of *sumak kawsay* (translated as '*buen vivir*' in Spanish), initially promoted in Ecuador's new constitution in 2008, when it became the first country in the world to codify 'the rights of nature', recognising the inalienable rights of ecosystems to exist and flourish. *Buen vivir* is rooted in a specifically anti-colonial, anti-capitalist world view, deeply critical of consumerist and individualistic interpretations of wealth and wellbeing, emphasising instead the importance of living in harmony with the cycles of Mother Earth (or 'Pachamama'). This has now spread far more widely through South America, though it has to be acknowledged that although much has been promised in countries like Ecuador, Colombia and Bolivia, a lot less has been delivered in practice.

In our Western cultures, the importance of connecting to nature in this way is all too often diminished by the seductive trappings of consumerism, by endless distractions, by lives lived digitally rather than physically. In her moving account (*Reclaiming the Wild Soul*) of how she overcame alcoholism and rediscovered 'the sunlight of the spirit' through nature, Mary Reynolds Thompson draws on the Greek myth about Erysichthon, King of Thessaly, who asked his men to cut down an oak tree sacred to the goddess Demeter. They refuse, because they know the tree grows on hallowed ground. So Erysichthon cuts it down himself. Demeter puts a curse on him: the more he eats, the hungrier he will grow. In the end, he consumes all he possesses, including his own children, and eventually his own flesh. She uses this myth to demonstrate the dangers of treating the Earth simply as a bank of resources to drive our growth-based economies, and shares with her readers the power of finding a very different path:

> There is a wild and creative energy that flows through the universe, and that is our birthright. We don't just live on the Earth; the Earth lives in us. As we open to the wisdom and wonder of the natural world, we discover that we are neither infallible gods who control the planet, nor are we merely small, helpless creatures. We are, above all, a force of nature. This is what our souls long for: to be part of this amazing world, connected and belonging.

RECONNECTING WITH NATURE

There's a growing realisation that *explicitly* addressing this crisis of disconnection (often referred to as 'nature deficit disorder', which I mentioned at the start of Chapter 13) is becoming more and more urgent – in practice as well as spiritually. Some of the earliest academic work on this was done in the UK by The Conservation Volunteers (an organisation of which I'm proud to be the president) with its extraordinary Green Gyms programme, demonstrating the substantial physical and mental health benefits experienced by people involved in conservation schemes of one kind or another. In October 2019, researchers at Leeds Beckett University analysed the social value of conservation projects run by the Wildlife Trusts offering volunteering opportunities to people experiencing mental health problems. For every £1 invested in such schemes, the social return on that investment was

£6.88, primarily in terms of reduced health costs, fewer visits to GPs and an easier return to work – leading to calls for all GPs to be able to prescribe participation in such activities as part of a new 'Natural Health Service'.

Much of this goes back to the ground-breaking work of the US author Richard Louv. His 2005 best-seller, *Last Child in the Woods: Saving our Children from Nature Deficit Disorder*, continues to resonate with many professionals working with young people today:

> Reducing that deficit – healing the broken bond between our young and Nature – is in our self-interest, not only because aesthetics or justice demands it, but also because our mental, physical and spiritual health depends upon it. The health of the Earth is at stake as well.

One immediate consequence of our new awareness about today's climate crisis is the growing number of educationists and professional bodies, in both the UK and the US, calling for changes in the school curriculum to reflect this current reality. Right now, for instance, there's an impressive campaign building to introduce a new GCSE in Natural History in the UK, with support from a leading exam board. But many initiatives in this space are simply tweaking today's subject-specific curriculum, which in itself is a reflection of our view of education seen primarily as a way of training young people for the world of work that awaits them – in a physical world apparently untouched by the emergencies of climate change and ecosystem collapse. Those assumptions are now so utterly detached from the reality of the lives those young people will be living in the future as to make any notion of conventional curriculum reform sound ridiculous. Might it not be time to think much more radically about how best to structure the education that young people need to overcome the disconnection between them and the natural world – by reconnecting education to the ways in which nature works? A fascinating new project (www.theharmonyproject.org.uk) has seized this particular bull by the horns in suggesting that the seven underlying principles of harmony in nature (including interdependence, diversity, adaptation and oneness) should be used as the building blocks of any 'fit for purpose' curriculum in these troubled times.

We have to think so much more deeply about enabling young people to find ways of living in this disruptive world – emotionally, psychologically and spiritually. I suspect more and more countries *will* now ensure that climate change is better reflected in the formal curriculum, but that's not the same thing. For instance, what educational resources will young people need to 'manage' the horror story of 3 billion wild creatures killed in Australia's bushfires? Or 'cope' with what will soon become the non-stop bombardment of one climate-induced disaster after another, with all their individual tragedies and collective grief?

Ensuring that as many young people as possible enjoy the healing benefits of engaging with nature – 'rewilding our souls', as George Monbiot puts it – will only take us so far. Beyond that, how can we prevent the blighting and the burning out of the human spirit in the face of such trauma? How can we address the underlying ideological causes of what's going on without exacerbating today's deeply divisive tribalism? Can we, in the words of climate justice activist Mary Annaïse Heglar, find 'a love strong enough to break through the terror [...] hot enough to shake you out of your despair and propel you to the front of the battlefield?'

I'm still amazed by the lengths to which people will go to depoliticise the Climate Emergency. Even as devastating an analysis of the climate crisis as that of David Wallace-Wells in *The Uninhabitable Earth* – so overwhelming, so terrifying, and so far beyond the limited understanding that most politicians have of the true nature of the climate crisis – still fails to draw a set of conclusions that his own logic would appear to be leading the reader towards. He pulls back from recommending *anything* that might really move the needle – indeed, from anything even remotely political. So too do most of the authors providing similarly searing accounts of just how bad things are out there, and just how much worse they're going to get in the short term, not just in the second half of the century. In his book *McMindfulness*, Professor Ronald Purser refers to a kind of 'quietist surrender' on the part of those mindfully contemplating the abuses of consumption-driven capitalism, and I can't help but think that there's a similar kind of surrender going on in the world of climate change.

This chapter's been all about acknowledging and internalising the gamut of emotions that 'telling the truth' about climate change inevitably elicits in more and more people. But facing up to those feelings is in no way a substitute for political action; indeed, it's the best psychological and spiritual resource we have in preparing ourselves for the fight ahead.

21

HOPE IN DISOBEDIENCE

*'You can't just sit around waiting for hope to come.
You don't seem to understand that hope is something
that you have to earn.'*

GRETA THUNBERG

If we'd started to get serious about reducing greenhouse gas emissions twenty years ago, at a time when we already had more than enough evidence of what would happen if we didn't, and had committed then to an annual reduction in greenhouse gases of around 4 per cent per annum, and then stuck with that 4 per cent reduction year after year, for the past twenty years, we would be on track right now to restrict that all-important temperature increase to below 1.5°C by the end of the century. But we didn't. In fact, we did the opposite. We kept on increasing emissions of greenhouse gases, year on year, for twenty years. As a direct consequence of that, starting from where we are now in 2020, we will need to reduce emissions by roughly *twice as much* every year, for the next twenty years.

Set against that reality, it's impossible not to be downcast by the patently inadequate response of world leaders today, especially given that almost all the solutions we need have been available to us for most of those 20 years. Yes, there's the Paris Agreement from 2015, which provides at least some sort of aspirational ambition level ('doing everything in our power' to stay below that 1.5°C threshold) and some sort of baseline for countries to implement their own decarbonisation strategies. However, bearing in mind that the best we can currently hope for – *if* countries successfully deliver on their Paris commitments – is an average temperature increase in excess of 3°C by the end of the century, then we're still in a very bad place indeed.

This grim conclusion was amply confirmed by the UN Climate Change Conference in Madrid at the end of 2019. It was widely acknowledged to have been the worst climate conference since the Conference in Copenhagen in 2009 (which collapsed in complete failure) – at precisely the time when the world needed a decisive and appropriately ambitious response to ever-worsening evidence of climate breakdown. At the time of going to press, things are definitely looking better. America is back in the game, and many countries are coming forward with more ambitious Net Zero targets. But there's still nothing that indicates we're on track for staying below that 2°C threshold, let alone 1.5°C.

We can take some comfort in the fact that it's not just governments and political parties that are now going to have to step up. As we saw back in Chapter 5, there's much to be encouraged about in terms of business and investor responses. It's harder for businesses and investors to go on denying reality than it is for politicians, for whom ideology all too often trumps rationality. All businesses get there in the end – witness the US Chamber of Commerce, one of the most outrageous pro-fossil fuel, denialist organisations in America, and one of the few to back President Trump's decision in 2017 to withdraw from the Paris Agreement. In November 2019, without so much as a 'sorry, guys, we really screwed up', it quietly changed the relevant page on its website so that it now reads: 'Greater collaboration between governments and businesses is essential to build the best models to tackle climate challenges, which is why the Chamber supports U.S. participation in the Paris Agreement.'

More importantly, when it comes to civil society, we constantly underestimate the strength of the climate movement and the speed with which things are changing; embedded in the whole emergency story is the idea of *emergence* – of new ideas and perspectives in people who've never thought much about this before; of new organisations positively seizing the moment; of new collective energy; of new artistic expressions of concern; of new and improbable alliances. In this regard, I've always loved Rebecca Solnit's 'mushroom analogy':

After a rain, mushrooms appear on the surface of the Earth as if from nowhere. Many do so from a sometimes vast underground fungus that remains invisible and largely unknown. What we call mushrooms, mycologists call the fruiting body of the larger, less visible fungus. Uprisings and revolutions are often considered to be spontaneous, but less visible long-term organising and groundwork – or underground work – often lay the foundations.

What's more, bit by bit, people are beginning to realise there's something much more profound going on here than simply 'solving the climate crisis'. You don't have to get very deep into a conversation about climate change without opening up wider considerations of how we're living our lives, our hopes and fears, our anxieties about the future, of today's economic insecurity. It's as if the Climate Emergency is giving us all permission to open up whole areas of our lives that often stay hidden – all the more important after COVID-19.

But given the disturbing and rapidly accelerating changes in the Earth's natural systems, is that really the best we're able to summon up by way of a collective human response? Set the one against the other, seen as a race between shifts in the climate and shifts in human behaviour, and you can't help but come to the conclusion not just that we're losing this race, but we don't even appear to know what it would take to win it.

As we've seen, it's this combination of the immensity of the challenge and the cumulative inadequacy of the response that has driven so many people to different variations on the theme of 'too late', to a continuum of emotional responses from resignation to despair. Interestingly, almost every 'too-later' I know, or whose work I've read, is seriously technophobic. They distrust technology; they have only bad stories to tell about technology (and don't mention life-enhancing things like vaccination or antibiotics whatever you do!); and they can see literally 'no hope' in anything that technology might make possible to keep the otherwise inevitable apocalypse at bay. For some, it could almost come across as if they're looking forward to the collapse they predict, so the last thing they want is some technofix getting in the way of that cauterising moment of truth. In their eyes, too many environmentalists suffer from debilitating techno-optimism, allowing us to ignore deeper problems about the nature of economic growth and our unsustainably wasteful lifestyles. I find this kind of technophobic fatalism deeply unhelpful, and to suggest that it's simply not possible to combine enthusiasm for technology with a deeper, more radical critique of our current politics and economy, is just foolish.

As far as the millions of young people who took part in the climate strikes in March and September 2019 are concerned, the only possible response to those who assert that it's already 'too late' has to be: 'How dare you?' Having been party, directly or indirectly, to at least three decades of indifference and empty rhetoric (studded with endless ironic references to the need 'to act on behalf of our children and our grandchildren'), more and more people are now telling them that it's too late to do anything about it. It doesn't play well.

Entitling his book *What We're Fighting for Now Is Each Other*, Wen Stephenson takes a very different approach:

Is it too late? We know what the science says, but what does your conscience say? What does 'too late' even mean? Too late for what? Even in the face of all we now know, will it ever be too late for some kind of faith in human decency; or to hold on to some kind of hope, however irrational it may seem, in our fellow human beings; or to love our brothers and sisters on this Earth? These things – faith and hope and love – are every bit as real as the science. As Martin Luther King knew, they are the very stuff that movements are made of: real movements, the kind of radical, transformative movements that have changed the course of history in the past, and maybe, just maybe, might change it again. If enough of us are willing to fight, and to fight hard enough, and to fight lovingly enough, and never give up. If we're willing to engage in this struggle – this radical and loving struggle – for each other.

I spent a lot of time in both March and September 2019 listening to the voices of young people. It was both humbling and disturbing to be exposed to their pain, to their lived experience of a world already profoundly disrupted by accelerating climate change, with things unravelling around them in a way that leaves many increasingly anxious about their future – as we saw in the preceding chapter. There were indeed notes of anger in the mix, but mostly regretful rather than recriminatory, surprisingly forgiving of an entire generation having been more or less absent without leave at the time when it mattered most. I suspect they will be less forgiving in the future, asking how it was that so many of us were only too happy to put off until tomorrow what needed to be so urgently addressed today. That's going to be a tough charge sheet to handle.

For more and more adults, it's literally impossible not to find a more appropriate response to those concerns, especially those whose children are just coming to terms with the reality of what this will mean *for the rest of their lives*. 'Tell the Truth' is indeed the only way to counter the kind of false hopes that have been deployed so effectively to obscure the science of climate change, even though we know that this can be psychologically devastating – for adults as well as for young people.

It's one of the reasons that makes Greta Thunberg's story so powerful, in that her activism today was fashioned in part by four years of intense grief and a feeling of helplessness as she learned more and more about climate change. As I explored back in Chapter 2, one route out of this was for her to persuade her parents to start thinking a lot more about their own lifestyle; as Greta herself says, she knew these changes, in and of themselves, would make next to no difference to the build-up of greenhouse gases in the atmosphere. But knowledge has consequences. When quizzed today about what lifestyle changes are most important, she urges adults to do their homework, to find out what's *really* going on, to dig down deep into the science of climate change.

I think that's the first place where adults need to go, setting time aside to do justice to what the science is really telling us. But as we get smarter in working out what it is that we're actually fighting for, it's not just our lifestyles that will need to change, but our world views. 'Solidarity' and 'justice' may not be words that pepper the average family's regular conversations, but dramatically cutting back on the amount of meat we eat, or choosing not to have a car, or to avoid flying whenever possible, are all actions consistent with these big themes, a necessary precondition for everything else we need to do – including our direct involvement in politics and climate campaigning.

Coming to terms with personal responsibility in that way can be testing, as pointed out by American writer Jonathan

Safran Foer: 'There is a far more pernicious form of science denial than Trump's: the form that parades as acceptance. Those of us who know what is happening, but do far too little about it are more deserving of the anger. We should be terrified of ourselves. We're the ones we have to defy. Self-recognition does not always indicate self-awareness. I am the person endangering my children.'

CIVIL DISOBEDIENCE

Every single year we delay, the harder it gets. A continuation of inadequate incrementalism over the next decade, let alone over the next thirty years, means only one thing: *it will by then be too late.* We will by then, for sure, be experiencing that grim progression from increasingly severe disruption in many parts of the world, to extreme disruption across the entire world, to the inevitable collapse of our civilisation. That's what scientists mean when they talk about 'having only ten to fifteen years left'. It's not that we have to get absolutely everything done within that very short period of time, but we do have to start *now* to have any chance at all of getting to a Net Zero economy within the next twenty years or so. Regrettably, most mainstream politicians do not, as yet, share that analysis. A few talk as if they do, but they don't really.

To all of which there is only one logical conclusion given where we are now: we have no alternative but to commit to more radical political action. To get as many people as possible involved in campaigning activities just as often as possible. To bring such pressure to bear on our political systems, while we still have time, to shift from today's wholly inadequate incrementalism to a full-on emergency response. The case for civil disobedience is now overwhelming. Just as the suffragettes found that they had no alternative but to be 'disorderly', all else having failed, many climate campaigners find they have no alternative but to be disobedient, all else having failed.

What shape that takes, and how campaigns evolve, will depend on different conditions in different countries. Whatever else happens, it's clear that the fossil fuel companies will be the target for a lot of that campaigning activity, particularly in the US. Their role in obscuring and confusing the science of climate change, in directly attacking climate scientists, in successfully obstructing the emergence of sensible policies from both the Republicans and the Democrats, and in corrupting the entire body politic in America through the use of billions of dollars of shareholders' money to prevent any serious action, has been felt more damagingly in the US than in any other country.

Having been reminded by Michael Mann's *The New Climate War* of the intensity of hostilities from the late 1980s onwards, it's hard not to see the directors of these fossil fuel companies, going back at least thirty years, as out-and-out climate criminals.

Mann tracks the origins of these battles to the setting up of the Global Climate Coalition (GCC) in 1989, as a direct riposte to the establishment of the Intergovernmental Panel on Climate Change in 1988. The GCC involved all major oil and gas companies, working hand in glove with billionaire funders like the Koch brothers or the Mercer family, often disseminating their toxic misrepresentations of the science of climate change through a web of right-wing think tanks.

The GCC was wound up in 2001, but hostilities carried on. Indeed, Mann cautions us very strongly against assuming that the war is now over, with many of those old fossil fuel combatants having morphed their nefarious activities into a different set of battle tactics:

> Fossil fuel interests, and those doing their bidding, have a single goal: inaction. We might henceforth call them inactivists. They come in various forms. The most hardcore contingent – the deniers – are in the process of going extinct (though there is still a remnant population of them). They are being replaced by other breeds of deceivers and dissemblers, namely, downplayers, deflectors, dividers, delayers and doomers – all willing participants in a multipronged strategy seeking to deflect blame, divide the public, delay action by promoting 'alternative' solutions that don't actually solve the problem, or insist we simply accept our fate – it's too late to do anything about it anyway, so we might as well keep the oil flowing. The climate wars have thus not ended. They have simply evolved into a new climate war.

I used to wonder how it was possible for employees of those companies (and particularly senior employees of BP and Shell with whom Forum for the Future worked as a 'critical friend' for more than 15 years) to square their knowledge of the science of climate change with their own moral compass. Those senior managers knew, as an irrefutable fact, that their basic business model threatened both the stability of the global economy and the longer-term prospects of humankind as a whole. Once knowledge of that kind has been internalised, for any individual, however well-meaning and sincere they may be, it must get harder and harder to look oneself in the mirror every morning and feel anything other than deep moral regret.

It's more nuanced today. Both these companies have started on their respective decarbonisation journeys – BP more purposefully than Shell. For all that, however, they are still powerful companies that make the lion's share of their profits from further exacerbating today's Climate Emergency. If they were, in any way, struck by Greta Thunberg's words in 2019 ('Act as if our house is on fire, because it is'), they must have seen themselves busily pouring petrol on those flames even as others set about extinguishing them.

I rather doubt that many employees, let alone directors, of these fossil fuel companies will be reading these words. They tend to avoid books of this kind. But if they haven't found it somewhere in their conscience, by now, at this very last moment, to get out of the oil and gas sector, it would be sensible for them to prepare for a decade of unremitting disruption.

There's still an important debate to be had about whether non-violent direct action (or civil disobedience of other kinds) actually *works* – in terms of securing change that would not otherwise have happened. There's always a balance between the tactical rationale for civil disobedience and the moral case that can be made regardless of tactical efficacy. Many of those arrested in the protests in 2019 were eloquent about the need 'to draw a clear moral line'; civil disobedience forces people to think more deeply about the tensions that exist in any society, and asks them where they personally will draw the line in their judgement of what is right and wrong.

There are still many countries that have not yet witnessed such civil disobedience at any scale, and where direct action is all but impossible. But it's clear to me that civil disobedience, used as a tactic to force politicians to respond in ways that are genuinely commensurate with today's Climate Emergency, is here to stay. It will take many different shapes and forms, with many different organisations stepping up to provide people with the means to stand up and be counted, in one way or another. Let's remember Solnit's mushroom analogy: the mushrooms come and go; underground, the fungus will continue to prosper.

As far as young people are concerned, I see no reason why the anger of the young should not continue to grow and grow as the full extent of our betrayal of them is revealed. Nearly 30 per cent of today's global population is under the age of eighteen, increasingly well-informed about climate change, increasingly connected through the internet. Today's school strikes will evolve in all sorts of ways that are impossible to predict at this stage. The phenomenon of 'intergenerational rage' will become one of the most significant factors of the next decade, not least as that kind of rage will be so important in sustaining young people's hopes.

When people talk about 'civil disobedience', they usually think of direct action protests, occupations and so on. But when schoolchildren go on strike, this too is a form of civil disobedience, deliberately but responsibly breaking the law that says all children below a certain age must be in school at times determined by lawmakers. And why would anybody think that the experience of disobedience at this young age will not turn into more and more engaged patterns of disobedience as young people make their way in a world increasingly disrupted by climate change?

In October 2019, huge numbers of adults joined the demonstrations organised by young people in order to show their solidarity; it's easy to imagine how this source of supportive adult activism will continue to grow. As the first President of the Czech Republic, Václav Havel, said:

> Either we have hope within us or we don't; it is a dimension of the soul; it's not essentially dependent on some particular observation of the world or estimate of the situation. Hope is not prognostication. It is an orientation of the spirit, an orientation of the heart; it transcends the world that is immediately experienced, and is anchored somewhere beyond its horizon.

I appreciate that *Hope in Hell* primarily offers a Western perspective, most relevant perhaps to Europe, the US, South America, Australia and New Zealand. But it's important to take into account the numbers of young people already involved elsewhere, with actions of one kind or another taken in more than 120 countries in 2019. What's more, all the projections tell us that the impacts of accelerating climate change are likely to be much graver in Asia, South-east Asia, the Middle East, Central America and the Caribbean. That will heighten the degree to which politicians are compelled to come up with more appropriate responses, with public concern regarding climate change often at least as high if not higher in countries like India, China and Brazil than it is in Europe and the US.

That doesn't make the politics of it any less difficult. Spare a thought, for instance, for young climate campaigners in Russia, the last major country to ratify the Paris Agreement in 2019. Draconian restrictions on public demonstrations have made it extremely hard for the Russian branch of Fridays for Future to establish a presence on the streets of Moscow and other major cities – even legal 'one person protests' have to maintain a gap between people of no less than 50 metres! So, in September 2019, with around 4 million people protesting in cities all around the world, there were just eighty-five protestors on the streets of Moscow. But the 'Greta effect' is at work here as it is in so many places: inspired by her 'go it alone' campaign back in 2018, wonderful young people like Arshak Makichyan and Margarita Naumenko are successfully binding Russia into a global movement – with or without Vladimir Putin's permission.

Protest in China is equally difficult. But the Chinese government has already shown itself to be surprisingly vulnerable to protests around air quality; despite the usual repressive tactics, especially regarding the internet and social media, there is still a highly sophisticated network of organisations and local groups making life very uncomfortable for local Chinese politicians.

And then there's Hong Kong, which witnessed such extraordinary scenes after the pro-democracy protests erupted in April 2019. The courage of protestors, almost 50 per cent of whom were in their twenties, weekend after weekend, was utterly remarkable.

Nothing has made me think more about the price that people are prepared to pay to defend what they hold most dear – in this case, the right to enjoy at least some of the entitlements of a democratic system of government. The protestors all understood just how precarious Hong Kong's democratic prospects were, living in the shadow of one of the most ruthlessly autocratic governments in the world. But even those slim hopes have now been crushed. Under cover of COVID-19, with many Western governments in disarray, particularly the USA, the Chinese moved to crush what was left of the pro-democracy movement, expelling four Opposition lawmakers in Hong Kong's Legislative Council at the end of November 2020, ensuring the resignation of the remaining fifteen. Beijing's takeover of Hong Kong was all but complete at that point, and the persecution of pro-democracy protesters has continued through 2021.

Such setbacks are painful. And they're also frequent, whether we're talking about democracy, social justice, racial equality, or climate change. However stubbornly hopeful we may be, we have to acknowledge that things will often get worse (probably a lot worse) before they get better.

Having come to terms with the fact that we do now find ourselves in that proverbial 'last chance saloon' when it comes to avoiding irreversible climate change, it makes no sense to try and persuade oneself that things are not as bad as they appear. At which point, it's impossible not to ask oneself the follow-up question of what sort of price one's prepared to pay, personally and professionally, in order to make a fitting contribution to turning things around.

I've never been all that comfortable laying down the law about what people should be doing in response to the challenges we now face; given my own cumulative carbon footprint, preachy lifestyle advice seems inappropriate. As I've said before, lifestyle shaming can be surprisingly counter-productive. But I always hope that some of the most obvious stuff just jumps off pages like these: buy things that last and look after them; eat less meat; put yourself empathetically in the shoes of others; support local shops and organisations, etc.

But as the quote from Greta Thunberg at the top of this chapter demonstrates, hope has to be earned. *In practice.* Mostly through lots and lots of little things – in not turning away from the reality of what is happening, in 'telling the truth' in conversations with family, friends and work col-leagues, in standing up for people who are standing up for us and for a better world, in challenging the intolerance and contempt of those for whom today's Climate Emergency remains a complete irrelevance. The best place to find hope is within yourself.

Beyond that, there's our time and our money. Hope can't be built on the back of someone else's actions. One of the most inspiring things going on today is the resurgence of local envi-ronmental initiatives, often as part of national networks such as Transition Towns, the Wildlife Trusts, The Conservation Volunteers, Friends of the Earth and so on, but mostly local as in 'uniquely of *this* place', responding directly to the needs and interests of a particular community or cause. Even in the busiest of lives there's time to help make things better through shared endeavour, empowering ourselves by empowering others. As many have said, the best way of earning hope is rolling up our sleeves.

The 'money bit' often gets neglected. But every time we buy something, or donate something, or invest in something (for instance, through a stakeholder pension), we're making a choice, and those choices can be seen as down payments on hope earned. There's a powerful generational story here that I believe will become more and more significant through this decisive decade, as those with lives already lived well start thinking more about the prospects of those just starting out in life. In those reassuring words that are used to define what is meant by 'sustainable development' ('development that meets the needs of the present without compromising the ability of future generations to meet their own needs') lies a mandate – a demand for intergenerational justice – that must now be honoured by older people.

That is what is now 'emergent' in my own life – as *the* focus for the work I intend to do from now on, rather than just part of the broader context. I'm not entirely sure what that will mean in practice, but it already feels closer to my time in the Green Party and Friends of the Earth in the 1970s and 1980s than it does to anything I've done since.

As I mentioned earlier, I've always supported the use of NVDA, through the Green Party and during my time at Friends of the Earth, where we were responsible for a number of actions that ended in some of our members getting arrested. I was never arrested myself. After leaving Friends of the Earth in 1991, my professional life took a very different turn, through Forum for the Future, with the Prince of Wales's Business and Sustainability Programme, as chair of the Sustainable Development Commission, on the board for nine years of the South West Regional Development Agency, as Chancellor of Keele University, as a non-executive director of Willmott Dixon or Wessex Water, as president of this and patron of that. I never felt compromised in terms of the work I was doing throughout that time, never felt that I was 'living a lie' in terms of the organisations I worked with, or that I was in any way downplaying the advice I've been giving. But for me personally, as someone who has spent the past twenty-five years working 'on the inside' to build a broader consensus for radical change, that's no longer appropriate as I come to terms with what will be asked of all of us in this decisive decade.

I'm *not* saying that this is true for everyone, but for someone like me to be 'hedging my bets' – by appearing to deny the implications of the science if not the science itself, by tolerating a level of risk as to the wellbeing of future generations that we know is literally intolerable – lacks integrity. In those circumstances, I can see no alternative but to help build a global movement in which more radical campaigning, including mass civil disobedience, will play a crucial part.

People will draw their own conclusions from this. Each of us has a different part to play in today's Climate Emergency. There's no right or wrong about any individual course of action. But what we *all* know now, irrefutably, is that this is literally the last decade in which authentic, grounded hope will be available to anchor everything we can do to serve our families, friends and future generations. Who knows what lies beyond this decade? But if we haven't dramatically changed our ways by then, genuine hope will have become the scarcest resource on Earth.

Resources

BOOKS

Mark Everard, *Rebuilding the Earth* (Palgrave Macmillan, 2020).

Paul Hawken, *Drawdown* (Penguin, 2018).

Rob Hopkins, *From What Is to What If: Unleashing the Power of Imagination to Create the Future We Want* (Chelsea Green Publishing, 2019).

Tim Jackson, *Prosperity Without Growth: Foundations for the Economy of Tomorrow* (Routledge, 2017).

Dahr Jamail, *The End of Ice: Bearing Witness and Finding Meaning in the Path of Climate Disruption* (The New Press, 2019).

Naomi Klein, *No Is Not Enough: Defeating the New Shock Politics* (Penguin, 2018).

Naomi Klein, *This Changes Everything: Capitalism vs. the Climate* (Penguin, 2015).

Michael McCarthy, *The Moth Snowstorm* (John Murray, 2016).

Bill McKibben, *Falter: Has the Human Game Begun to Play Out?* (Henry Holt & Co, 2019).

Joanna Macy and Chris Johnston, *Active Hope* (New World Library, 2012).

Jane Mayer, *Dark Money* (Scribe Publications, 2016).

Donella H. Meadows, Jorgen Randers, Dennis L. Meadows, *Limits to Growth: The 30-Year Update* (Chelsea Green Publishing, 2004).

George Monbiot, *Out of the Wreckage: A New Politics for an Age of Crisis* (Verso, 2017).

Kristin Ohlson, *The Soil Will Save Us: How Scientists, Farmers and Foodies Are Healing the Soil to Save the Planet* (Rodale Books, 2014).

Jonathon Porritt, *The World We Made: Alex McKay's Story from 2050* (Phaidon Press, 2013).

Kate Raworth, *Doughnut Economics: Seven Ways to Think Like a 21st-Century Economist* (Random House Business, 2018).

Rebecca Solnit, *Hope in the Dark* (Canongate Books, 2016).

Wen Stephenson, *What We're Fighting for Now Is Each Other: Dispatches from the Front Lines of Climate Justice* (Beacon Press, 2015).

Greta Thunberg, *No One Is Too Small to Make a Difference* (Penguin, 2019).

Greta Thunberg, Malena Ernman, Beata Ernman, Svante Thunberg, *Our House Is on Fire* (Allen Lane, 2020).

Peter Wadhams, *A Farewell to Ice: A Report from the Arctic* (Allen Lane, 2016).

David Wallace-Wells, *The Uninhabitable Earth* (Penguin Random House, 2013).

ORGANISATIONS

350.org: https://350.org/

Bloomberg New Energy Finance: https://about.bnef.com/

Cambridge Institute for Sustainability Leadership (CISL): https://www.cisl.cam.ac.uk/

Carbon Tax Center: https://www.carbontax.org/you-us/who-we-are/

Centre for Alternative Technology: https://www.cat.org.uk/

ClientEarth: https://www.clientearth.org/

Climate Accountability Institute: https://climateaccountability.org/

The Climate Group: https://www.theclimategroup.org/

The Climate Mobilization: https://www.theclimatemobilization.org/

Compassion in World Farming (CiWF): https://www.ciwf.org.uk/

Energy Transitions Commission: http://www.energy-transitions.org/

Extinction Rebellion: https://rebellion.earth/

Fair Tax Mark: https://fairtaxmark.net/

Flight Free UK: https://flightfree.co.uk/

The Food and Land Use Coalition (FOLU): https://www.foodandlanduse coalition.org/

Forum for the Future: https://www.forumforthefuture.org/

Fridays for Future: https://www.fridaysforfuture.org/

Friends of the Earth International: https://www.foei.org/

Friends of the Earth UK: https://friendsoftheearth.uk/

Global Financial Integrity: https://gfintegrity.org/

Global Witness: https://www.globalwitness.org/en/

Grantham Institute: https://www.imperial.ac.uk/grantham/

Green New Deal Group: https://greennewdealgroup.org/

Green New Deal UK: https://www.greennewdealuk.org/

The Green Party: https://vote.greenparty.org.uk/

Greenpeace International: https://www.greenpeace.org/international/

Greenpeace UK: https://www.greenpeace.org.uk/

InfluenceMap: https://influencemap.org/

The Intergovernmental Panel on Climate Change (IPCC): https://www.ipcc.ch/

Intergovernmental Science-Policy Platform on Biodiversity and Ecosystem Services (IPBES): https://ipbes.net/

International Union for Conservation of Nature (IUCN): https://www.iucn.org/

Make Votes Matter: https://www.makevotesmatter.org.uk/

Natural Resources Defense Council (NRDC): https://www.nrdc.org/

New Economics Foundation (NEF): https://neweconomics.org/

Population Matters: https://populationmatters.org/

Possible: https://www.wearepossible.org/

Project Drawdown: https://drawdown.org/

Reboot the Future: https://www.rebootthefuture.org/

RethinkX: https://www.rethinkx.com/

The Royal Society: https://royalsociety.org/

Scott Polar Research Institute: https://www.spri.cam.ac.uk/

SolarAid: https://solar-aid.org/

SOS UK: https://sustainability.nus.org.uk/

Stop Ecocide: https://www.stopecocide.earth/

Sunrise Movement: https://www.sunrisemovement.org/

Survival International: https://www.survivalinternational.org/

Sustainable Energy for All (SEforALL): https://www.seforall.org/

The Sustainable Food Trust: https://sustainablefoodtrust.org/

Task Force on Climate-related Financial Disclosures (TCFD): https://www.fsb-tcfd.org/

Tax Justice Network: https://www.taxjustice.net/

The Terraton Initiative: https://www.terraton.org/

UK Student Climate Network: https://ukscn.org/

Union of Concerned Scientists: https://www.ucsusa.org/

World Values Survey: http://www.worldvaluessurvey.org/wvs.jsp

REFERENCES

CHAPTER 1: THIS IS PERSONAL

7   *The Paris Agreement*
     United Nations, 'Paris Agreement', United Nations Framework
     Convention on Climate Change, 2015; downloadable from:
     https://unfccc.int/process-and-meetings/the-paris-agreement/
     the-paris-agreement

CHAPTER 2: THE POWER OF HOPE

12   *'Hope is an embrace of the unknown'*
     Rebecca Solnit, *Hope in the Dark* (Canongate Books, 2016).
13   *'the negativity instinct'*
     Hans Rosling, Ola Rosling and Anna Rosling Rönnlund,
     Factfulness: Ten Reasons We're Wrong About the World – and Why
     Things Are Better Than You Think (Sceptre, 2018).
15   *'Act like your house is on fire. Because it is.'*
     Greta Thunberg, Speech to the World Economic
     Forum, 20 September 2019: https://www.youtube.com/
     watch?v=U72xkMz6Pxk (at 5 minutes 35 seconds)
15   *'For those of us who are on the spectrum'*
     Greta Thunberg, TEDxStockholm, November
     2018: https://www.youtube.com/watch?time_
     continue=19&v=EAmmUIEsN9A&feature=emb_logo (at 2 minutes
     10 seconds)
16   *'Why should we be studying for a future'*
     Greta Thunberg, Speech to UN Secretary General António
     Guterres, Katowice, 3 December 2018: https://www.youtube.com/
     watch?time_continue=2&v=Hq489387cg4&feature=emb_logo (at
     2 minutes 10 seconds)

17   *'Sudden shifts can happen'*
Joanna Macy and Chris Johnston, *Active Hope* (New World Library, 2012).

19   *'the human rights struggle of our time'*
Naomi Klein, quoted in Wen Stephenson, *What We're Fighting for Now Is Each Other: Dispatches from the Front Lines of Climate Justice* (Beacon Press, 2015); for the Klein quote, see: https://earthbound.report/2019/05/20/what-were-fighting-for-now-is-each-other-by-wen-stephenson/

19   *'I'm a human being.'*
Stephenson, *What We're Fighting for Now Is Each Other*, op. cit.

21   *'Hope is not something that you have.'*
Alexandria Ocasio-Cortez, quoted in Emma Brockes, 'When Alexandria Ocasio-Cortez met Greta Thunberg', *The Guardian*, 29 June 2019: https://www.theguardian.com/environment/2019/jun/29/alexandria-ocasio-cortez-met-greta-thunberg-hope-contagious-climate

22   *'I meet more people every day who have given up'*
Rob Hopkins, *From What Is to What If: Unleashing the Power of Imagination to Create the Future We Want* (Chelsea Green Publishing, 2019).

22   *'I take some comfort from the work of US political scientist Erica Chenoweth'*
Erica Chenoweth, 'It May Only Take 3.5% of the Population to Topple a Dictator – With Civil Resistance', *The Guardian*, 1 February 2018: https://www.theguardian.com/commentisfree/2017/feb/01/worried-american-democracy-study-activist-techniques

CHAPTER 3: MAKING SENSE OF THE SCIENCE

27   *'In October 2018, the IPCC brought out a Special Report ("Global Warming of 1.5°C")'*
V. Masson-Delmotte, P. Zhai, *et al.* (eds), 'Global Warming of 1.5°C: An IPCC Special Report on the impacts of global warming of 1.5°C above pre-industrial levels and related global greenhouse gas emission pathways, in the context of strengthening the global response to the threat of climate change, sustainable development, and efforts to eradicate poverty' (IPCC, 2018); online version: https://www.ipcc.ch/sr15/about/; download access: https://www.ipcc.ch/sr15/download/#full

31   *'concluded that between a quarter and a third of the emission reductions required'*

Bronson W. Griscom *et al.*, 'Natural Climate Solutions', *Proceedings of the National Academy of Sciences*, 114/44 (31 October 2017): 11645–50; https://doi.org/10.1073/pnas.1710465114

31   *'If we're serious about climate change'*
Mark Tercek, quoted in Alister Doyle, 'Plant More Trees to Combat Climate Change: Scientists', Reuters, 16 October 2017: https://www.reuters.com/article/us-climate-change-land/plant-more-trees-to-combat-climate-change-scientists-idUSKBN1CL2PP

32   *'We have to do everything, and we have to do it immediately.'*
Graham Lawton, quoting Piers Forster, 'Hitting 1.5°C', *New Scientist*, 240/3207 (8 December 2018): https://www.sciencedirect.com/science/article/pii/S0262407918322590; https://doi.org/10.1016/S0262-4079(18)32259-0

34   *'a record figure of 59.1 billion tonnes of CO_2e'*
United Nations Environment Programme, 'Emissions Gap Report 2019', p. 3; downloadable from: https://www.unenvironment.org/resources/emissions-gap-report-2019

35   *'The last time the Earth experienced a comparable concentration of CO_2'*
World Meteorological Organization, 'Greenhouse Gas Levels in Atmosphere Reach New Record', Press Release 22112018, 20 November 2018 (quoting Petteri Taalas): https://public.wmo.int/en/media/press-release/greenhouse-gas-levels-atmosphere-reach-new-record

38   *'We are on track for a temperature rise of over 3°C'*
United Nations Environment Programme, tweet by @UNEP, 26 November 2019: https://twitter.com/UNEP/status/1199333716255895552

CHAPTER 4: THE POWER OF TECHNOLOGY

39   *'an appalling delusion'*
Damian Carrington, 'Idea of renewables powering UK is an "appalling delusion"' (quoting David MacKay), *The Guardian*, 3 May 2016: https://www.theguardian.com/environment/2016/may/03/idea-of-renewables-powering-uk-is-an-appalling-delusion-david-mackay

44   *'The IEA has also estimated that China's offshore wind capacity'*
International Energy Agency, 'Offshore Wind to Become a $1 Trillion Industry', press release, 25 October 2019: https://www.iea.org/news/offshore-wind-to-become-a-1-trillion-industry

45   *Table: Contribution of renewable energy to five large industrial economies.*

International Energy Authority, 2019 data except the 'World' column, 2018: https://www.iea.org/

45   *'Germany will soon get around half of its power from renewables'*
David Elliott, 'The EU is Doing Well on Green Power – as the UK Exits', blog, Renew Extra Weekly, 1 February 2020: https://renewextraweekly.blogspot.com/2020/02/there-have-been-divergent-views-on.html

47   *'In its so-called "Sustainable Development Scenario"'*
International Energy Agency, 'World Energy Model: Scenario Analysis of Future Energy Trends', November 2019: https://www.iea.org/reports/world-energy-model/sustainable-development-scenario

48   *'The most authoritative study for the UK'*
Centre for Alternative Technology, 'Zero Carbon Britain: Rising to the Climate Emergency', 2019; downloadable from: https://www.cat.org.uk/info-resources/zero-carbon-britain/research-reports/zero-carbon-britain-rising-to-the-climate-emergency/; report PDF: https://www.cat.org.uk/download/35541/

48   *'Even today, more than ten times as many people are employed'*
Adam Vaughan, 'US Green Economy has 10 Times More Jobs than the Fossil Fuel Industry', *New Scientist*, Issue 3252, 19 October 2019: https://www.newscientist.com/article/2219927-us-green-economy-has-10-times-more-jobs-than-the-fossil-fuel-industry/

50   *'We are on the cusp of one of the fastest, deepest, most consequential disruptions'*
James Arbib and Tony Seba, 'Rethinking Transportation 2020–2030: The Disruption of Transportation and the Collapse of the Internal-Combustion Vehicle and Oil Industries', RethinkX, May 2017: https://www.rethinkx.com/transportation

51   *'In the words of GM's CEO, Mary Barra'*
Coral Davenport, 'G.M. Drops Its Support for Trump Climate Rollbacks and Aligns with Biden', *The New York Times*, 23 November 2020: https://www.nytimes.com/2020/11/23/climate/general-motors-trump.html

51   *'EVs will likely start to erode the last major bastion of oil demand growth'*
Bank of America Merrill Lynch, in Alex Longley,
'BofA Sees Oil Demand Peaking by 2030 as Electric Vehicles Boom', Bloomberg, 22 January 2018: https://www.bloomberg.com/news/articles/2018-01-22/bofa-sees-oil-demand-peaking-by-2030-as-electric-vehicles-boom

51   *'When it comes to raw materials there is simply no comparison.'*
Lucien Mathieu, quoted in Jasper Jolly, 'Fossil fuel cars make

'hundreds of times' more waste than electric cars', *Guardian*, 1 March 2021.

51   *'According to Bloomberg New Energy Finance, the "tipping point"'*
Bloomberg NEF, 'Electric Vehicle Outlook 2019', 15 May 2019, Bloomberg Finance L.P. 2019: https://www.eenews.net/assets/2019/05/15/document_ew_02.pdf

52   *'At this juncture, we will likely see the beginning of the collapse'*
Jeremy Rifkin, *The Green New Deal: Why the Fossil Fuel Civilization Will Collapse by 2028, and the Bold Economic Plan to Save Life on Earth* (New York: St Martin's Press, 2019).

52   *'Between 2010 and 2018, SUVs doubled their market share'*
Niko Kommenda, 'SUVs Second Biggest Cause of Emissions Rise, Figures Reveal', *The Guardian*, 25 October 2019: https://www.theguardian.com/environment/ng-interactive/2019/oct/25/suvs-second-biggest-cause-of-emissions-rise-figures-reveal

53   *'Delhi is the world's most polluted city'*
Robin Hicks, 'Air Pollution is Killing Thousands More People Than Covid in the World's Biggest Cities', *Eco-Business*, 18 February 2021.

54   *'A hard-hitting report from the Business and Human Rights Resource Centre'*
Tim Ha, 'Exploitation Rife Among Firms Mining Minerals for Renewables, Electric Vehicles: Report', Eco-Business, 5 September 2019: https://www.eco-business.com/news/exploitation-rife-among-firms-mining-minerals-for-renewables-electric-vehicles-report/

54   *'A new report from Amnesty International in February 2021'*
Amnesty International, 'Powering Change: Principles for Businesses and Governments in the Battery Value Chain', 4 February 2021: https://www.amnesty.org/en/documents/act30/3544/2021/en/

55   *'According to the 2019 Global Trends in Renewable Energy Investment Report'*
Frankfurt School-UNEP Centre/BNEF, 'Global Trends in Renewable Energy Investment 2019': http://www.fs-unep-centre.org; PDF: http://wedocs.unep.org/bitstream/handle/20.500.11822/29752/GTR2019.pdf

57   *'In the UK, the Green New Deal Group is pushing this "jobs in every constituency"'*
https://greennewdealgroup.org/jobs-in-every-constituency-the-promise-of-the-green-new-deal/

58   *'I used to scoff at the notion that using power more efficiently '*

Bill Gates, *How to Avoid a Climate Disaster: The Solutions We Have and the Breakthroughs We Need* (Allen Lane, 2021).

61   '*In November 2020, the latest report from asset management company Lazard*' https://www.lazard.com/perspective/levelized-cost-of-energy-and-levelized-cost-of-storage-2020/

CHAPTER 5: THE GREAT AWAKENING

63   '*According to Media Matters for America,* total *coverage of climate change issues*'
      Kevin Kalhoefer, 'How Broadcast Networks Covered Climate Change in 2016', Media Matters for America, 23 March 2017: https://www.mediamatters.org/donald-trump/how-broadcast-networks-covered-climate-change-2016

64   '*28 per cent of citizens questioned considered climate change to be "a crisis"*'
      Jennifer de Pinto, Fred Backus and Anthony Salvanto, 'Most Americans Say Climate Change Should be Addressed Now – CBS News Poll', CBS News, 15 September 2019: https://www.cbsnews.com/news/cbs-news-poll-most-americans-say-climate-change-should-be-addressed-now-2019-09-15/

64   '*77 per cent of younger Republicans saw climate change as a serious threat*'
      Andrew Rafferty and Ellen Rolfes, 'How Young Conservatives Hope to Make Climate a GOP Issue', Newsy, 4 September 2019; data referred to is in video discussing IPSOS/Newsy survey results, starting at 3 minutes 50 seconds: https://www.newsy.com/stories/how-conservatives-are-trying-to-make-climate-a-gop-issue/

65   '*I will be on the Capitol every Friday for the next four months*'
      Jane Fonda, 'Fire Drill Fridays', blog, Jane Fonda, 8 November 2019: https://www.janefonda.com/2019/11/fire-drill-fridays/

65   '*You need to know that your stand is our stand.*'
      Matthew Taylor, quoting Sharan Burrow, 'Trade Unions Around the World Support Global Climate Strike', *The Guardian*, 19 September 2019: https://www.theguardian.com/environment/2019/sep/19/trade-unions-around-the-world-support-global-climate-strike

66   '*This is all wrong.*'
      Greta Thunberg, Speech to UN Climate Action Summit, 23 September 2019, reported in UN News, 23 September 2019; audio: https://news.un.org/en/story/2019/09/1047052; video: https://www.youtube.com/watch?v=OGVShq47C4o

67   '*AOC's video,* A Message From the Future.'
      Naomi Klein, 'A Message from the Future', video, *The Intercept with Alexandria Ocasio-Cortez*, 17

April 2019: https://theintercept.com/2019/04/17/
green-new-deal-short-film-alexandria-ocasio-cortez/

68   'Data gathered by Eurobarometer shows that nearly 80 per cent of EU
      citizens surveyed'
      Kantar Belgium, 'Special Eurobarometer 490 Report: Climate
      Change', European Commission, April 2019; downloadable from:
      https://ec.europa.eu/commfrontoffice/publicopinion/index.
      cfm/survey/getsurveydetail/instruments/special/surveyky/2212;
      doi:10.2834/00469

68   "at 71 per cent, India has the highest percentage of people in any country'
      Matthew Smith, 'International Poll: Most Expect to
      Feel Impact of Climate Change, Many Think It Will
      Make Us Extinct', YouGov, 15 September 2019: https://
      yougov.co.uk/topics/science/articles-reports/2019/09/15/
      international-poll-most-expect-feel-impact-climate

69   'Its update in July 2020 confirmed that about 53 per cent'
      https://eciu.net/netzerotracker

69   'Our house is on fire. The European Parliament has seen the blaze'
      Greenpeace European Unit, 'Declaring a Climate
      Emergency Is Not Enough – Our House Is On Fire',
      press release, 28 November 2019: https://www.
      greenpeace.org/eu-unit/issues/climate-energy/2490/
      european-parliament-declaring-climate-emergency-not-enough/

71   'But at the time of writing there are more than 1,400 companies'
      Science Based Targets, 'Companies Taking Action', webpage,
      undated: https://sciencebasedtargets.org/companies-taking-action/

71   'At the time of writing, there are now 300 global companies signed up'
      For current list of RE100 Companies see https://www.there100.
      org/re100-members

72   'By 2030, our ambition is to reduce more greenhouse gas emissions than are
      emitted'
      IKEA News, 'IKEA Accelerates Movement to Become Climate
      Positive, Committing to Reduce Absolute Greenhouse Gas
      Emissions from Production by 80%', 30 November 2018:
      https://newsroom.inter.ikea.com/news/
      ikea-accelerates-movement-to-become-climate-
      positive-committing-to-reduce-absolute-greenhouse-
      gas-e/s/45bfcf5a-742b-4bde-8744-5ca9de43c0b8

72   'the top twenty companies involved in extracting and selling fossil fuels'
      Matthew Taylor and Jonathan Watts, 'Revealed: The 20 Firms
      Behind a Third of All Carbon Emissions', The Guardian, 9 October
      2019: https://www.theguardian.com/environment/2019/oct/09/
      revealed-20-firms-third-carbon-emissions

74   *'We have reached the inflection point where, in some cases'*
        'Lazard Released Annual Levelized Cost of Energy and Levelized
        Cost of Storage Analyses', Lazard Ltd, press release, 8 November 2018:
        https://lazardltd.gcs-web.com/news-releases/news-release-details/
        lazard-releases-annual-levelized-cost-energy-and-levelized-0

74   *'The vast majority of fossil fuel reserves are "unburnable"'*
        Jessica Shankleman for *BusinessGreen*, quoting Mark Carney, 'Mark
        Carney: Most Fossil Fuel Reserves Can't Be Burned', *The Guardian*,
        13 October 2014: https://www.theguardian.com/environment/2014/
        oct/13/mark-carney-fossil-fuel-reserves-burned-carbon-bubble

75   *'As of April 2021, support for the TCFD had grown to more than 2,000
        organisations'*
        See https://www.fsb-tcfd.org/support-tcfd/

75   *'a devastating report from Rainforest Action Network'*
        'Banking on Climate Change: Fossil Fuel Finance Report Card
        2019', Rainforest Action Network, BankTrack, Indigenous
        Environmental Network (IEN), Sierra Club, Oil Change
        International, and Honor the Earth, March 2019; downloadable
        from: https://www.ran.org/bankingonclimatechange2019/

75   *'Companies that don't adapt will go bankrupt'*
        Damian Carrington, quoting Mark Carney, 'Firms Ignoring
        Climate Crisis Will Go Bankrupt', *The Guardian*, 13 October
        2019: https://www.theguardian.com/environment/2019/oct/13/
        firms-ignoring-climate-crisis-bankrupt-mark-carney-bank-
        england-governor

76   *'The latest update on the* GoFossilFree.org *website records more than
        1,200 divestment commitments'*
        https://gofossilfree.org/divestment/commitments/

76   *'Research has shown that investments that score well in terms of ESG
        credentials'*
        Nigel Green, quoted in '8 out of 10 millennials now prioritise
        responsible investing – and they're right', press release, deVere
        Group, 2 January 2020: https://www.devere-group.com/
        news/8-out-of-10-millennials-now-prioritise-responsible-
        investing-and-theyre-right.aspx

77   *'Bloomberg New Energy Finance announced in October 2019'*
        Bloomberg New Energy Finance, 'Sustainable Debt Joins the
        Trillion-Dollar Club', blog, 17 October 2019: https://about.bnef.
        com/blog/sustainable-debt-joins-the-trillion-dollar-club/

79   *'Exercising my "reasoned judgment", I have no doubt'*
        Judge Ann Aiken, *Opinion and Order* in the United States District
        Court for the District of Oregon, Eugene Division, Juliana *et al.* v.
        United States of America *et al.*, Case Number 6:15-cv-01517-TC,

Document 83, 10 November 2016, p. 32: https://static1.
squarespace.com/static/571d109b04426270152febe0/t/5824e85e6a
49638292ddd1c9/1478813795912/Order+MTD.Aiken.pdf

79 *'the plaintiffs have made a compelling case that action is needed'*
Andrew D. Hurwitz, *Opinion* in the United States Court
of Appeals for the Ninth Circuit, Juliana *et al.* v. United States
of America *et al.*, Case Number 18–36082, ID: 11565804,
DktEntry 153-1, p. 31; downloadable from: http://climatecase
chart.com/case/juliana-v-united-states

80 *'Estimates by the Pacific Institute'*
Fred Pearce, 'Polluter Pays?', *New Scientist*, 239/3191 (18 August
2018), pp. 38–41: https://www.sciencedirect.com/science/
article/abs/pii/S0262407918314842; https://doi.org/10.1016/
S0262-4079(18)31484-2

83 *'Jim Hansen has consistently complained of "scientific reticence"'*
James Hansen, 'Scientific Reticence and Sea Level
Rise', *Environmental Research Letters*, 2/2, 24 May
2007; downloadable from: https://iopscience.iop.org/
article/10.1088/1748-9326/2/2/024002; PDF: https://
iopscience.iop.org/article/10.1088/1748-9326/2/2/024002/pdf;
doi:10.1088/1748-9326/2/2/024002

83 *'with too many scientists "erring on the side of least drama"'*
Keynyn Bryssem, Naomi Oreskes, *et al.*, 'Climate change
prediction: Erring on the side of least drama?', *Global Environmental
Change*, 23 (2013), pp. 327–37; PDF: http://www.phys.uri.edu/
nigh/FFRI/LeastDrama.pdf

84 *'as unsolvable, feeding a sense of doom, inevitability and hopelessness'*
Michael E. Mann, Facebook post, 10 July 2017: https://www.
facebook.com/MichaelMannScientist/posts/1470539096335621

84 *'Climate doomism can be paralysing'*
Michael E. Mann, *The New Climate War: The Fight to Take Back our
Planet* (Scribe UK, 2021).

86 *'But the significant success of David Wallace-Wells's* The Uninhabitable
Earth*'*
David Wallace-Wells, *The Uninhabitable Earth* (Penguin Random
House, 2013).

87 *'The survey found that people's readiness to accept that direct link'*
As reported in Jennifer Marlon and Abigail Cheskis, 'Wildfires
and Climate Are Related – Are Americans Connecting the
Dots?', blog, Yale Program on Climate Change Communication,

11 December 2017: https://climatecommunication.yale.edu/
news-events/connecting-wildfires-with-climate/

87   *'as Jim Hansen reminds us'*
James Hansen, 'Why I Must Speak Out About Climate Change',
TED2012, talk/video, February 2012: https://www.ted.com/talks/
james_hansen_why_i_must_speak_out_about_climate_change/
transcript?language=en

88   *'The World Health Organization (WHO) estimates that heatwaves
already kill'*
Quoted in Alan Miller, Renilde Becque, *et al.*, 'Chilling Prospects:
Providing Sustainable Cooling for All', Sustainable Energy for All
(SEforALL), 1 July 2018, p. 20; downloadable from: https://www.
seforall.org/publications/chilling-prospects-cooling-for-all-report;
PDF: https://www.seforall.org/sites/default/files/SEforALL_
CoolingForAll-Report.pdf

89   *'all of these heatwaves, on average, are "twice as likely" to have happened'*
University of Oxford, 'Heatwave made "twice as likely by climate
change"', online article, 27 July 2018: http://www.ox.ac.uk/
news/2018-07-27-heatwave-made-twice-likely-climate-change#

89   *'suggested that by 2050 London was likely to have the climate of
Barcelona'*
Jean-Francois Bastin, Emily Clark, *et al.*, 'Understanding climate
change from a global analysis of city analogues', *PLOS One*, 14(10):
e0224120 (2019): https://doi.org/10.1371/journal.pone.0217592;
https://journals.plos.org/plosone/article?id=10.1371/journal.
pone.0217592

89   *'Overall, by 2050, it's been estimated that losses in productivity'*
Tord Kjellstrom *et al.*, 'Threats to Occupational Health and Labour
Productivity in the Economy from Increasing Heat During
Climate Change', ClimateCHIP, Technical Report 2014: 2,
15 December 2014: http://www.climatechip.org/sites/default/
files/publications/Technical%20Report%202_Climate%20
change%2C%20Workplace%20Heat%20exposure%2C%20
Health%2C%20%20Labor%20Productivity%2C%20and%20
the%20Economy.pdf

92   *'That's bad enough, but the work of Professor Irakli Loladze reveals that'*
Irakli Loladze, 'Hidden Shift of the Ionome of Plants Exposed to
Elevated CO_2 Depletes Minerals at the Base of Human Nutrition',
eLife (2014): https://elifesciences.org/articles/02245; doi: 10.7554/
eLife.02245

CHAPTER 7: MELTING ICE AND RISING SEAS

96   *'the AMOC is now weaker than it has ever been'*
      L. Caesar, G. D. McCarthy et al, 'AMOC Weakest in Last
      Millennium' *Nature Geoscience* 14, 118–120 (2021).
      https://www.nature.com/articles/s41561-021-00699-z

97   *'Every fraction of a degree of warming avoided'*
      Adam Vaughan, 'Greenland Can't Go Green', *New Scientist*,
      242/3228, 4 May 2019, p. 23: https://www.sciencedirect.com/
      science/article/pii/S0262407919307808

98   *'It revealed that the rate of melting in West Antarctica had accelerated
      threefold'*
      S. R. Rintoul, S. L. Chown, *et al.*, 'Choosing the Future
      of Antarctica', *Nature* 558, pp. 233–241 (2018): https://doi.
      org/10.1038/s41586-018-0173-4; https://www.nature.com/
      articles/s41586-018-0173-4

99   *'Some of the changes Antarctica will face are already irreversible'*
      Hayley Dunning, quoting Martin Siegert, 'How to Save
      Antarctica – and the Rest of Earth Too', blog, Imperial College
      London, 13 June 2019: https://www.imperial.ac.uk/news/186668/
      how-save-antarctica-rest- earth/

100   *'But a new report from NASA in December 2018 showed that'*
      Maria-José Viñas, 'More Glaciers in East Antarctica Are Waking
      Up', NASA, 10 December 2018: https://www.nasa.gov/feature/
      goddard/2018/more-glaciers-in-antarctica-are-waking-up

101   *'A report published in* Science Advances *in 2019 showed that Himalayan
      glaciers'*
      J. M. Maurer, J. M. Schaefer, *et al.*, 'Acceleration of Ice Loss Across
      the Himalayas Over the Past 40 Years', *Science Advances*, 5/6 (2019):
      https://advances.sciencemag.org/content/5/6/eaav7266.full

103   *'suggested we could be seeing a 3-metre sea-level rise – not by 2100 but by
      2050.'*
      James Hansen, Makiko Sato, *et al.*, 'Ice Melt, Sea Level Rise and
      Superstorms: Evidence from Paleoclimate Data, Climate Modelling,
      and Modern Observations that 2°C Global Warming Could Be
      Dangerous', *Atmospheric Chemistry and Physics*, 16 (2016), 3761–812:
      https://www.atmos-chem-phys.net/16/3761/2016/acp-16-3761-2016.
      pdf

103   *'14,000 years ago, as the Ice Age began to loosen its grip'*
      Bill McKibben, *Falter: Has the Human Game Begun to Play Out?*
      (Henry Holt & Co, 2019).

104   *'As the tideline rises higher than the ground people call home'*
      Jonathan Watts, quoting Scott Kulp, 'Rising sea levels pose threat

to homes of 300m people – study', *The Guardian*, 29 October 2019: https://www.theguardian.com/environment/2019/oct/29/rising-sea-levels-pose-threat-to-homes-of-300m-people-study

CHAPTER 8: FEEDBACK LOOPS AND TIPPING POINTS

107   *'and if there's one research area for which he's best known, it's the state of sea ice'*
        Peter Wadhams, *A Farewell to Ice: A Report from the Arctic* (Allen Lane, 2016).

107   *'In the end, it will just melt away quite suddenly.'*
        Jonathan Amos, quoting Peter Wadhams, 'Arctic Summers Ice-Free "by 2013"', BBC, 12 December 2007: http://news.bbc.co.uk/1/hi/sci/tech/7139797.stm

107   *'Every year, however, the US National Oceanic and Atmospheric Administration'*
        National Oceanic and Atmospheric Administration, 'Arctic Report Card 2018', p. 2; downloadable from: https://arctic.noaa.gov/Report-Card/Report-Card-Archive

108   *'The latest models are basically showing that no matter'*
        Quoted in Gloria Dickie, 'The Arctic is in a death spiral. How much longer will it last?', *the Guardian* 13 October 2020.

109   *'Experts at the Scripps Institution of Oceanography believe'*
        Chris Mooney, 'The Arctic Ocean has lost 95 percent of its oldest ice – a startling sign of what's to come', e-newsletter, *Other News* (2018): https://mailchi.mp/other-net/the-arctic-ocean-has-lost-95-percent-of-its-oldest-ice-a-startling-sign-of-whats-to-come

109   *'The report fails to focus on the weakest link in the climate chain'*
        Fiona Harvey, quoting Durwood Zaelke, '"Tipping Points" Could Exacerbate Climate Crisis, Scientists Fear', *The Guardian*, 9 October 2018: https://www.theguardian.com/environment/2018/oct/09/tipping-points-could-exacerbate-climate-crisis-scientists-fear

109   *'Some 17 per cent of the rainforest has already been entirely destroyed'*
        Timothy M. Lenton, Johan Rockström, *et al.*, 'Climate Tipping Points – Too Risky to Bet Against', *Nature*, 575 (2019): 592–5; doi: 10.1038/d41586-019-03595-0; https://www.nature.com/articles/d41586-019-03595-0

110   *'an article in the journal Nature suggested that we may have already crossed a series of tipping points'*
        Timothy M. Lenton, Johan Rockström, et al, 'Climate Tipping Points – Too Risky to Bet Against', *Nature*, 27 November 2019

(as corrected 9 April 2020): https://www.nature.com/articles/
d41586-019-03595-0#correction-0

111   *'Arctic sea ice retreat, directly induced by greenhouse gas warming'*
Wadhams, *A Farewell to Ice*, op. cit.

114   *'I fear it is a collective failure of nerve by those whose responsibility
it is'*
Wadhams, *A Farewell to Ice*, op. cit.

116   *'In 2012, the indomitable Stephen Salter'*
Environmental Audit Committee, 'Protecting the Arctic: Further
Written Evidence from Professor Stephen Salter, Emeritus
Professor of Engineering Design, Edinburgh University', UK
Parliament, 5 March 2012: https://publications.parliament.uk/pa/
cm201213/cmselect/cmenvaud/171/171we13.htm

CHAPTER 9: OUT OF TIME? JUST IN TIME?

121   *'He offers us a disturbing end-of-world continuum'*
Jem Bendell, 'Deep Adaptation: A Map for Navigating Climate
Tragedy', *IFLAS* Occasional Paper 2, 27 July 2018: http://
lifeworth.com/deepadaptation.pdf

122   *'In their book,* This Civilisation is Finished'
Rupert Read and Samuel Alexander, *This Civilisation is Finished*
(Simplicity Institute, 2019)

123   *'There is now no weather we haven't touched'*
Kate Marvel, 'We Need Courage, Not Hope, to Face Climate
Change', blog, On Being, 1 March 2018: https://onbeing.org/
blog/kate-marvel-we-need-courage-not-hope-to-face-climate-
change/

124   *'It's* not *too late.'*
Oliver Milman, quoting Jim Hansen, 'Ex-Nasa Scientist:
30 Years on, World is Failing "Miserably" to Address
Climate Change', *The Guardian*, 19 June 2018: https://
www.theguardian.com/environment/2018/jun/19/
james-hansen-nasa-scientist-climate-change-warning

126   *'any progressive cause needs no more than around 3.5 per cent of the population'*
Erica Chenoweth, 'My Talk at TEDxBoulder: Civil
Resistance and the "3.5% Rule"', Rationalinsurgent, 4
November 2013: https://rationalinsurgent.com/2013/11/04/
my-talk-at-tedxboulder-civil-resistance-and-the-3-5-rule/

128   *'These clashes [. . .] feel like an early warning'*
Gaby Hinsliff, 'Extinction Rebellion Has Built Up So Much
Goodwill. It Mustn't Throw That Away', *The Guardian*,
17 October 2019 (amended 18 October 2019): https://

www.theguardian.com/commentisfree/2019/oct/17/
extinction-rebellion-canning-town-well-off-people

130   *'Remember, there is no "too late" here'*
David Roberts, 'Hope and Fellowship', Grist, 30 August 2013:
https://grist.org/climate-energy/hope-and-fellowship/

CHAPTER 10: NARRATIVES OF HOPE

135   *'At the time of going to press, more than 1,900 jurisdictions and
municipalities'*
https://www.cedamia.org/global/

137   *'We must understand that the forces of denial'*
Mann, *The New Climate War*, op. cit.

139   *'Owing to past neglect, in the face of the plainest warnings'*
Winston Churchill, 'Debate on the Address', 1117, HC Deb,
12 November 1936, Vol. 317, cc1081–155: https://api.parliament.uk/
historic-hansard/commons/1936/nov/12/debate-on-the-address

140   *'It's clear that society is capable of responding dramatically to major threats'*
Jørgen Randers and Paul Gilding, 'The One Degree War Plan',
2010; downloadable from https://paulgilding.com/2009/11/06/
cc20091106-odw-launch/

141   *'"victory plan for the Earth"'*
Ezra Silk, 'Victory Plan', The Climate Mobilization, March 2019;
downloadable from https://www.theclimatemobilization.org/
victory-plan

142   *'Climate breakdown has to be framed as a matter of justice'*
John Barry and Noel Healy, 'From Declarations of Climate
Emergencies to Climate Action: Narratives and Strategies of
Wartime and Citizen Mobilisations for Rapid and Just Transitions',
Queen's University Belfast, 8 August 2019; http://qpol.qub.ac.uk/
from-declarations-of-climate-emergencies-to-climate-action-
narratives-and-strategies-of-wartime-and-citizen-mobilisations-
for-rapid-and-just-transitions/

145   *'from introducing social security and minimum wage laws'*
Naomi Klein, *On Fire: The Burning Case for a Green New Deal*
(Allen Lane, 2019).

145   *'From its very beginning, the United States'*
Robinson Meyer, quoting Cohen and DeLong, 'A Centuries-Old
Idea Could Revolutionize Climate Policy', *The Atlantic*, 19 February
2019: https://www.theatlantic.com/science/archive/2019/02/green-
new-deal-economic-principles/582943/; Stephen S. Cohen and
J. Bradford DeLong, *Concrete Economics: The Hamilton Approach to
Economic Growth and Policy* (Harvard Business Review Press, 2016).

146 *'This wave of globalization has wiped out our middle class.'*
TIME, quoting Donald Trump, 'Read Donald Trump's
Speech on Trade', 28 June 2016: https://time.com/4386335/
donald-trump-trade-speech-transcript/
147 *'Here's the only way any of this works'*
David Roberts, 'This Is An Emergency, Damn It',
blog, Vox, 23 February 2019: https://www.vox.
com/energy-and-environment/2019/2/23/18228142/
green-new-deal-critics
147, 148 *'The most remarkable day in the history'*
'His orders set up or strengthen'
Bill McKibben, 'The Biden Administration's Landmark Day in the
Fight for the Climate', *New Yorker*, 28 January 2021.
153 *'finds no evidence that modern societies have managed to decouple'*
T. Parrique, J. Barth, *et al.*, 'Decoupling Debunked: Evidence
and Arguments Against Green Growth as a Sole Strategy for
Sustainability', European Environmental Bureau, 2019: https://eeb.
org/library/decoupling-debunked/
154 *'Every country's Green New Deal must make sure that we actually hit'*
Klein, *On Fire*, op. cit.

CHAPTER 11: THE HARD STUFF

155 *'At the back end of 2018, a high-powered international body'*
Energy Transitions Commission, 'Mission Possible: Reaching
Net-Zero Carbon Emissions from Harder-to-Abate Sectors by
Mid-Century', November 2018; downloadable from: http://www.
energy-transitions.org/sites/default/files/ETC_MissionPossible_
FullReport.pdf
156 *'One of the go-to gurus in this area, Professor Julian Allwood'*
Julian M. Allwood and Jonathan M. Cullen, *Sustainable Materials –
with Both Eyes Open: Future Buildings, Vehicles, Products and
Equipment – Made Efficiently and Made with Less New Material
(Without the Hot Air)* (UIT Cambridge, 2011).
157 *'Companies are constantly trying to drive down costs'*
ArcelorMittal, 'Climate Action Report 1 2018', May 2019;
downloadable from: https://corporate.arcelormittal.com/
sustainability/approach/climate-change; PDF: https://
storagearcelormittaluat.blob.core.windows.net/media/3lqlqwoo/
climate-action-report-2019.pdf
158 *'the building sector could shave off as much as a third of its carbon footprint'*
'The Circular Economy – a Powerful Force for Climate
Mitigation', *Material Economics Sverige AB*, Stockholm, 2018;

downloadable from: https://materialeconomics.com/publications/
the–circular–economy

164   'a company called LanzaTech'
        https://www.lanzatech.com/

166   'Why not speed up the process and make oil from algae living today?'
        Ruth Kassinger, Bloom: From Food to Fuel, The Epic Story of How
        Algae Can Save Our World (Elliott & Thompson Ltd, 2019).

169   'This is a historic moment'
        Guillaume Faury, quoted in 'Airbus reveals new Zero Emissions
        Concept Aircraft', www.Airbus.com, September 2020.
        https://www.airbus.com/newsroom/press-releases/en/2020/09/
        airbus-reveals-new-zeroemission-concept-aircraft.html

171   'a company called Carbon Engineering had successfully taken CO$_2$ out of the
        air'
        David W. Keith, Geoffrey Holmes, David St. Angelo and Kenton
        Heidel, Joule, 2, 1573–94, 15 August 2018, published by Elsevier
        Inc: https://doi.org/10.1016/j.joule.2018.05.006

173   'As explained on the Carbon Tax Center's excellent website'
        https://www.carbontax.org/

CHAPTER 12: WE HAVE TO TALK ABOUT GEOENGINEERING

177   'The deliberate large-scale manipulation of an environmental process'
        Working group, 'Geoengineering the Climate: Science, Governance
        and Uncertainty', The Royal Society, September 2009, p.
        1: https://royalsociety.org/topics-policy/publications/2009/
        geoengineering-climate/

179   'a worldwide tree-planting programme on more than 1.5 billion hectares'
        Damian Carrington, 'Tree planting "has mind-blowing
        potential" to tackle climate crisis', The Guardian, 4 July 2019:
        https://www.theguardian.com/environment/2019/jul/04/planting-
        billions-trees-best-tackle-climate-crisis-scientists-canopy-emissions;
        Introducing Jean-Francois Bastin, Yelena Finegold, et al., 'The
        Global Tree Restoration Potential', Science, 365/6448, 5 July 2019:
        https://science.sciencemag.org/content/365/6448/76

186   'The idea of "fixing" the climate by hacking the Earth's reflection of sunlight'
        Raymond T. Pierrehumbert, 'Climate Hacking is Barking Mad',
        Slate, 10 February 2015: https://slate.com/technology/2015/02/
        nrc-geoengineering-report-climate-hacking-is-dangerous-and-
        barking-mad.html

188   'Clive Hamilton lays bare just how devious and untrustworthy'
        Clive Hamilton, Earthmasters: The Dawn of the Age of Climate
        Engineering (Yale University Press, 2013).

CHAPTER 13: PEAK MEAT

192 *'Land use contributes about one-quarter of global greenhouse gas emissions'*
P. R. Shukla, J. Skea, E. Calvo Buendia, V. Masson–Delmotte,
H.-O. Pörtner, D. C. Roberts, P. Zhai, R. Slade, S. Connors, R.
van Diemen, M. Ferrat, E. Haughey, S. Luz, S. Neogi, M. Pathak,
J. Petzold, J. Portugal Pereira, P. Vyas, E. Huntley, K. Kissick, M.
Belkacemi, J. Malley (eds.), 'Climate Change and Land: an IPCC
special report on climate change, desertification, land degradation,
sustainable land management, food security, and greenhouse gas
fluxes in terrestrial ecosystems', IPCC, 2019.

193 *'According to the UN, 1.8 billion people will be living in countries or regions'*
https://www.un.org/waterforlifedecade/scarcity.shtml

194 *'Politicians are still saying'*
Tim Benton, Carling Bieg, et al, 'Food System Impacts on
Biodiversity Loss', Chatham House, 3 February 2021:
https://www.chathamhouse.org/2021/02/
food-system-impacts-biodiversity-loss

195 *'Around three per cent of US citizens are now vegans'*
https://news.gallup.com/poll/267074/percentage-americans-
vegetarian.aspx

196 *'There is a strong positive relationship between the level of income'*
World Health Organization, 'Nutrition. 3. Global and Regional
Food Consumption Patterns and Trends: 3.4 Availability and
Changes in Consumption of Animal Products', undated: https://
www.who.int/nutrition/topics/3_foodconsumption/en/index4.
html

196 *'more and more citizens are following a meat-free diet'*
UK Diet Trends 2021: https://www.finder.com/uk/uk-diet-trends

197 *'From the Latin for 'forest' and 'grazing', silvopasture'*
Paul Hawken, *Drawdown: The Most Comprehensive Plan Ever Proposed
to Reverse Global Warming* (Penguin, 2018).

198 *'governments are spending between $700 billion and $1 trillion'*
Food and Land Use Coalition, September 2019, p. 7:
https://www.theguardian.com/environment/2019/sep/16/1m-a-
minute-the-farming-subsidies-destroying-the-world;
https://www.foodandlandusecoalition.org/wp-content/
uploads/2019/09/FOLU-GrowingBetter-GlobalReport.pdf

199 *'In many countries, there is a worrying disconnect between the retail price of food'*
Food and Agriculture Organization of the United Nations,
'Natural Capital Impacts in Agriculture: Supporting Better

Business Decision-Making', June 2015, p. 5; downloadable from: http://www.fao.org/nr/sustainability/natural-capital

200   *'UK consumers spend £120 billion on food every year'*
Ian Fitzpatrick, Richard Young, Robert Barbour, *et al.*, 'The Hidden Cost of UK Food', Sustainable Food Trust, November 2017 (revised and corrected July 2019); downloadable from: https://sustainablefoodtrust.org/articles/hidden-cost-uk-food/; PDF: https://sustainablefoodtrust.org/wp-content/uploads/2013/04/Website-Version-The-Hidden-Cost-of-UK-Food.pdf

201   *'at least 2 million people in the US are infected with antibiotic-resistant bacteria'*
Centers for Disease Control and Prevention, 'Antibiotic Resistance Threats in the United States, 2019' (Atlanta, GA: US Department of Health and Human Services, CDC, 2019): downloadable from: https://www.cdc.gov/drugresistance/biggest-threats.html; PDF: https://www.cdc.gov/drugresistance/pdf/threats-report/2019-ar-threats-report-508.pdf

201   *'Antimicrobial resistance poses a catastrophic threat.'*
S. C. Davies, 'Annual Report of the Chief Medical Officer, Volume Two, 2011: Infections and the rise of antimicrobial resistance' (London: Department of Health, 2013): https://assets.publishing.service.gov.uk/government/uploads/system/uploads/attachment_data/file/138331/CMO_Annual_Report_Volume_2_2011.pdf

202   *'Unlike the cow, we get better and better at making meat every single day.'*
Dara Kerr, quoting Patrick Brown, 'Impossible Burger 2.0 Tastes Like Beef. Really', CNET, 7 January 2019: https://www.cnet.com/news/impossible-burger-2-0-tastes-like-beef-really/

204   *'We are on the cusp of the deepest, fastest, most consequential disruption'*
Catherine Tubb and Tony Seba, 'Rethinking Food and Agriculture 2020–2030: The Second Domestication of Plants and Animals, the Disruption of the Cow, and the Collapse of Industrial Livestock Farming', RethinkX, September 2019, p. 6; downloadable from: https://www.rethinkx.com/food-and-agriculture

205   *'If our grazing land was allowed to revert to natural ecosystems'*
George Monbiot, 'We Can't Keep Eating as We Are – Why Isn't the IPCC Shouting This from the Rooftops?', *The Guardian*, 8 August 2019: https://www.theguardian.com/commentisfree/2019/aug/08/ipcc-land-climate-report-carbon-cost-meat-dairy

CHAPTER 14: CHINA: HEADING FOR THE ABYSS

209   'There are ways to stop toxic pollution'
        Richard Smith, China's Engine of Environmental Collapse (Pluto
        Press, 2020).

210   'produced a powerful documentary called Under the Dome'
        Under the Dome, film written, produced and directed by Chai Jing
        (2015): https://filmsfortheearth.org/en/films/under-the-dome

CHAPTER 15: LETHAL INCUMBENCIES

226   'It claimed that the poverty rate had been cut in half since 2000.'
        United Nations, 'The Millennium Development Goals Report
        2015' (New York: United Nations, 2015), p. 15; PDF: https://
        www.un.org/millenniumgoals/2015_MDG_Report/pdf/
        MDG%202015%20rev%20%28July%201%29.pdf

226   'the US-based think-tank Global Financial Integrity revealed the true
        picture'
        Dev Kar, 'Financial Flows and Tax Havens: Combining to Limit
        the Lives of Billions of People', press release, Global Financial
        Integrity, 5 December 2016; introducing report of the same title
        by Dev Kar, Guttorm Schjelderup, et al. (Washington: Global
        Financial Integrity, December 2015): https://gfintegrity.org/
        report/financial-flows-and-tax-havens-combining-to-limit-the-
        lives-of-billions-of-people/

227   'The good news narrative serves as a potent political tool.'
        Jason Hickel, The Divide: Global Inequality from Conquest to Free
        Markets (W. W. Norton & Company, 2018).

228   'in 2017, Oxfam released its wealth report showing that the richest eight
        people'
        Larry Elliott, 'World's Eight Richest People Have Same Wealth
        as Poorest 50%', The Guardian, 16 January 2017: https://www.
        theguardian.com/global-development/2017/jan/16/worlds-eight-
        richest-people-have-same-wealth-as-poorest-50; 'Just 8 Men Own
        Same Wealth as Half the World', press release, Oxfam, 16 January
        2018: https://www.oxfam.org/en/press-releases/just-8-men-own-
        same-wealth-half-world; Introducing Oxfam's Briefing Paper, 'An
        Economy for the 99%': downloadable from https://www.oxfam.
        org/en/research/economy-99

229   'Immense wealth translates automatically into immense environmental
        impacts'
        George Monbiot, 'For the Sake of Life on Earth, We Must Put a

Limit on Wealth', *The Guardian,* 19 September 2019:
https://www.theguardian.com/commentisfree/2019/sep/19/life-earth-wealth-megarich-spending-power-environmental-damage

230   *'Globally, subsidies remained large at $4.7 trillion'*
David Coady, Ian Parry, *et al.*, 'Global Fossil Fuel Subsidies
Remain Large: An Update Based on Country Estimates',
IMF Working Paper No 19/89, 2 May 2019: https://
www.imf.org/en/Publications/WP/Issues/2019/05/02/
Global-Fossil-Fuel-Subsidies-Remain-Large-An-Update-Based-on-Country-Level-Estimates-46509

231   *'Their apparent prominence in the public sphere'*
Mann, *The New Climate War,* op. cit.

232   *'the five largest publicly owned oil and gas companies spend around $200
million'*
Niall McCarthy, 'Oil Firms Spend Millions on Climate Lobbying',
InfluenceMap, 26 March 2019: https://www.statista.com/
chart/17467/annual-expenditure-on-climate-lobbying-by-oil-and-gas-companies/

232   *'If you look at net investment in low-carbon alternatives'*
Ron Bousso, 'Big Oil Spent 1 percent on Green Energy in 2018',
Reuters, 12 November 2018: https://www.reuters.com/article/
us-oil-renewables/big-oil-spent-1-percent-on-green-energy-in-2018-idUSKCN1NH004
Chart with specific figures: https://fingfx.thomsonreuters.com/
gfx/ce/7/1800/1799/Pasted%20Image.jpg

233   *'the Climate Emergency meets financial contagion.'*
Paul Gilding, 'Climate Contagion: 2020–2025. So it Begins', blog,
The Cockatoo Chronicles, 13 December 2019: https://paulgilding.
com/2019/12/13/climate-contagion-2020-2025/

234   *'Back in 2006, Nicholas Stern's "Review of the Economics of Climate
Change"'*
Nicholas Stern, 'The Economics of Climate Change: The Stern
Review' (Cambridge University Press, 2007). (First published as
'The Stern Review on the Economic Effects of Climate Change'
in 2006, but no longer available in that edition.)

235   *'Every now and then, the United Nations Environment Programme produces'*
United Nations Environment Programme, 'Global Environment
Outlook 6', 2019: https://www.unenvironment.org/resources/
global-environment-outlook-6

236   *'Morgan Stanley reported that weather disasters "intensified by climate
change"'*
'Weather Disasters "Intensified by Climate Change" Cost North
America $650bn ...', reported in Tom DiChristopher, 'Climate

Disasters Cost the World $650 Billion Over 3 years – Americans are Bearing the Brunt: Morgan Stanley', CNBC, 14 February 2019: https://www.cnbc.com/2019/02/14/climate-disasters-cost-650-billion-over-3-years-morgan-stanley.html

236   *'If the risk from wildfires, flooding, storms or hail is increasing'* Arthur Neslen, quoting Ernst Rauch, 'Climate Change Could Make Insurance Too Expensive for Most People – Report', *The Guardian*, 21 March 2019: https://www.theguardian.com/environment/2019/mar/21/climate-change-could-make-insurance-too-expensive-for-ordinary-people-report

238   *'We may need to take major steps, perhaps including a world levy'* Thatcher MSS, Churchill Archive Centre: THCR 5/1/5/576 28 f9, online at the Margaret Thatcher Foundation: Thatcher MSS, 22 September 1988, 'Environment: Morris Minute for Guise (Lawson comments on second draft of Royal Society speech)': https://www.margaretthatcher.org/document/208608

239   *'The Chancellor has seen the draft speech.'* Thatcher MSS, ibid.

240   *'The late 1980s was the absolute zenith of the neoliberal crusade'* Naomi Klein, 'Capitalism Killed our Climate Momentum, not "Human Nature"', *The Intercept*, 3 August 2018: https://theintercept.com/2018/08/03/climate-change-new-york-times-magazine/

241   *'The conditions for success could not have been more favorable.'* Nathaniel Rich, *Losing Earth: The Decade We Could Have Stopped Climate Change* (Picador, 2019).

CHAPTER 16: DEMOCRACY AT RISK

242   *'It's obviously very hard to unearth the full extent of this corrupting influence'* Jane Mayer, *Dark Money* (Scribe Publications, 2016).

243   *'The UK with its corporate tax haven network'* 'UK and Territories are "Greatest Enabler" of Tax Avoidance, Study Says', *The Guardian*, 28 May 2019: https://www.theguardian.com/world/2019/may/28/uk-and-territories-are-greatest-enabler-of-tax-avoidance-study-says

224   *'Money flows across frontiers, but laws do not.'* Oliver Bullough, *Moneyland: Why Thieves and Crooks Now Rule the World and How to Take it Back* (Profile Books, 2018).

245   *'there are now significant minorities in both the USA and Europe'* Roberto Stefan Foa and Yascha Mounk, 'The Democratic Disconnect', *Journal of Democracy*, 27/3 (July 2016), p. 9: https://

www.journalofdemocracy.org/wp-content/uploads/2016/07/
FoaMounk-27-3.pdf

245   *'In January 2020, it concluded that only 430 million people'*
'The Retreat of Global Democracy Stopped in 2018: Or Has it Just
Paused?', *The Economist*, 8 January 2019: https://www.economist.
com/graphic-detail/2019/01/08/the-retreat-of-global-democracy-
stopped-in-2018; reporting on: 'Democracy Index', *The Economist
Intelligence Unit Limited* (2019): https://www.eiu.com/public/
topical_report.aspx?campaignid=democracyindex2019

245   *'almost a third of the world's people who now live in notionally democratic'*
V-Dem Institute, 'Democracy for All?': V-Dem Annual Democracy
Report 2018', University of Gothenburg (2018), p. 6: https://
www.v-dem.net/media/filer_public/68/51/685150f0-47e1-4d03-
97bc-45609c3f158d/v-dem_annual_dem_report_2018.pdf; https://
www.v-dem.net/media/filer_public/90/44/90444c35-f97e-44f1-
943e-320e55edfc8e/v-dem_democracy_report_2018.pdf

246   *'Asked if they agreed that "Britain needs a strong ruler willing to break the
rules"'*
Hansard Society, 'Audit of Political Engagement 16', 2019,
p. 51; downloadable from: https://www.hansardsociety.org.
uk/publications/reports/audit-of-political-engagement-16;
report in PowerPoint format: https://assets.ctfassets.
net/rdwvqctnt75b/7iQEHtrkIbLcrUkduGmo9b/
cb429a657e97cad61e61853c05c8c4d1/Hansard-Society__Audit-of-
Political-Engagement-16__2019-report.pdf

246   *'around a third of people in any country are psychologically predisposed'*
Karen Stenner, *The Authoritarian Dynamic* (Cambridge University
Press, 2007).

248   *'The continued use of extreme ideologies'*
Corey J. A. Bradshaw, Paul Ehrlich et al, *Frontiers in Conservation
Science*, 13 January 2021. https://www.frontiersin.org/
articles/10.3389/fcosc.2020.615419/full

248   *'coarsener-in-chief', 'a larger-than-life, over-the-top avatar of narcissism',
'Trump is a troll'*
Michiko Kakutani, *The Death of Truth* (William Collins, 2018).

249   *'In their important book How Democracies Die'*
Steven Levitsky and Daniel Ziblatt, *How Democracies Die: What
History Reveals About Our Future* (Viking, 2018).

251   *'The team estimates that far-right disinformation networks'*
Carl Miller, 'Inside the Information Wars', *New Scientist*, 244/3252
(19 October 2019): pp. 38–41: https://www.sciencedirect.
com/science/article/abs/pii/S0262407919319736; https://doi.
org/10.1016/S0262-4079(19)31973-6

252 'We do not believe it should be our role' (quoting Richard Allan).
'While internet advertising is incredibly powerful' (quoting Jack Dorsey).
Donna Lu, 'Fighting Fake News', New Scientist, 244/3255
(9 November 2019): pp. 3–56: https://www.sciencedirect.
com/science/article/abs/pii/S0262407919320895; https://doi.
org/10.1016/S0262-4079(19)32089-5

253 'Facebook has too much power over our economy, our society and our
democracy.'
Elizabeth Warren, tweet @ewarren, 12 May 2019: https://
twitter.com/ewarren/status/1127697561115033606; linking
to article by Elizabeth Warren, 'It's Time to Break up
Amazon, Google, and Facebook', blog, Team Warren,
8 March 2019: https://medium.com/@teamwarren/
heres-how-we-can-break-up-big-tech-9ad9e0da324c

253 'the big six US tech companies [. . .] have "aggressively avoided" $100bn of
tax'
Fair Tax Mark, 'Tax Gap of Silicon Six over $100 Billion so Far
this Decade', press release, 2 December 2019: https://fairtaxmark.
net/tax-gap-of-silicon-six-over-100-billion-so-far-this-decade/

254 'soaring rates of self-harm, eating disorders, anxiety and body dysmorphia'
William Storr, Selfie: How the West Became Self-Obsessed (Pan
Macmillan/Picador, 2018).

254 'If those who believe in the market economy and liberal democracy'
Martin Wolf, 'Why So Little has Changed Since the Financial
Crash', Financial Times, 4 September 2018: https://www.ft.com/
content/c85b9792-aad1-11e8-94bd-cba20d67390c

255 'asked in a poll in 2019 if it would be "a good thing for the army to take over"'
Foa and Mounk, 'The Democratic Disconnect' (2016), op. cit.

255 'We're at an inflection point'
Joseph Biden, 'Remarks by President Biden at the 2021
Virtual Munich Security Conference', the White House, 19
February 2021. https://www.whitehouse.gov/briefing-room/
speeches-remarks/2021/02/19/remarks-by-president-biden-at-the-
2021-virtual-munich-security-conference/

256 'The ideal subject of totalitarian rule is not the convinced Nazi'
Hannah Arendt, The Origins of Totalitarianism (London: Secker &
Warburg, 1951).

257 'We are all born into a world we did not make'
Astra Taylor, 'Bad Ancestors: Does the Climate Crisis Violate the
Rights of Those Yet to be Born?', The Guardian, 1 October 2019:
https://www.theguardian.com/environment/2019/oct/01/bad-
ancestors-climate-crisis-democracy; Astra Taylor, Democracy May
Not Exist, But We'll Miss It When It's Gone (Verso, 2019).

CHAPTER 17: PLANETARY PRESSURES AND OPPORTUNITIES

259 *'The health of the ecosystems on which we and other species depend'*
Bob Watson, quoted in Intergovernmental Science-Policy
Platform on Biodiversity and Ecosystem Services, media
release, IPBES Secretariat, 7 May 2019: https://ipbes.net/news/
Media-Release-Global-Assessment

260 *'There are now more than 700 sites worldwide affected by low oxygen
conditions'*
International Union for Conservation of Nature, 'Marine Life,
Fisheries Increasingly Threatened as the Ocean Loses Oxygen –
IUCN Report', December 2019: https://www.iucn.org/news/
marine-and-polar/201912/marine-life-fisheries-increasingly-
threatened-ocean-loses-oxygen-iucn-report; downloadable from:
https://portals.iucn.org/library/node/48892

261 *'the rate of extinction in insect species'*
Damian Carrington, quoting Francisco Sánchez-Bayo,
'Plummeting Insect Numbers "Threaten Collapse of Nature"',
The Guardian, 10 February 2019: https://www.theguardian.com/
environment/2019/feb/10/plummeting-insect-numbers-threaten-
collapse-of-nature; reporting on Francisco Sánchez-Bayo and
Kris A. G. Wyckhuys, 'Worldwide Decline of the Entomofauna:
A Review of its Drivers', *Biological Conservation*, 232 (April 2019),
pp. 8–27: https://www.sciencedirect.com/science/article/pii/
S0006320718313636?via%3Dihub; https://doi.org/10.1016/j.
biocon.2019.01.020

262 *'as much as 25 per cent of the Earth's productive land is now so degraded'*
Food and Agriculture Organization, 'The State of the World's
Land and Water Resources for Food and Agriculture: Managing
Systems at Risk', The Food and Agriculture Organization
of the United Nations and Earthscan (2011), p. 113:
http://www.fao.org/3/i1688e/i1688e.pdf; Global Environment
Facility, 'Land Degradation: Main Issue', undated webpage,
Global Environment Facility: https://www.thegef.org/topics/
land-degradation

262 *'Soil depth in Iowa (at the very heart of the US breadbasket)'*
Jeremy Grantham, 'The Race of Our Lives Revisited', GMO
White Paper, GMO LLC (August 2018), hosted on DivestInvest
website.

262 *'degradation of one kind or another costs the global economy between $6.3
trillion'*
ELD Initiative, 'The Value of Land: Prosperous lands and positive
rewards through sustainable land management', *The Economics of*

Land Degradation (2015), p. 51: https://www.eld-initiative.org/
fileadmin/pdf/ELD-main-report_05_web_72dpi.pdf

263   *'Weirdly, we've all been schooled in the notion that plants are takers'*
Kristin Ohlson, *The Soil Will Save Us: How Scientists, Farmers and
Foodies Are Healing the Soil to Save the Planet* (Rodale Books, 2014),
p. 232.

265   *'Growing human populations have significant implications'*
Partha Dasgupta, 'The Economics of Biodiversity:
The Dasgupta Review', *HM Treasury*, 2nd February
2021. https://www.gov.uk/government/publications/
final-report-the-economics-of-biodiversity-the-dasgupta-review

265   *'There is only one soil, one flora, one fauna, and one people.'*
Aldo Leopold, 'Land Pathology', Lecture to the University of
Wisconsin Chapter of Sigma Xi, 15 April 1935, in Susan L. Flader
and J. Baird Callicott (eds), *The River of the Mother of God and Other
Essays by Aldo Leopold* (University of Wisconsin Press, March 1991),
pp 212–17.

265   *'The human population is expected to grow to around 11 billion'*
United Nations Department of Economic and Social Affairs,
'World Population Projected to Reach 9.8 Billion in 2050, and 11.2
Billion in 2100', 21 June 2017: https://www.un.org/development/
desa/en/news/population/world-population-prospects-2017.html

266   *'Among its thirteen recommendations to end the destruction'*
As reported by Population Matters, '20,000 Scientists Ignored –
Time for Action', 11 July 2018: https://populationmatters.
org/news/2018/07/11/20000-scientists-ignored-time-action;
reporting on William J. Ripple, Christopher Wolf, *et al.*, 'World
Scientists' Warning to Humanity: A Second Notice', *Bioscience*,
67/12 (December 2017), pp. 1026–8: https://academic.oup.com/
bioscience/article/67/12/1026/4605229; https://doi.org/10.1093/
biosci/bix125

266   *'limited population growth, up from 6.6 billion in 2019 to 7.6 billion by 2050'*
UN World Population Prospects: September 2019: https://
population.un.org/wpp/Publications/Files/WPP2019_Highlights.
pdf

268   *'44 per cent of all pregnancies in the world today are unplanned'*
Nathan Hodson, 'Climate Change and Contraception', blog, *BMJ
Sexual & Reproductive Health*, 16 October 2019: https://blogs.bmj.
com/bmjsrh/2019/10/16/climate-change-and-contraception/;
introducing J. Bongaarts and R. Sitruk-Ware, 'Climate Change
and Contraception', *BMJ Sexual and Reproductive Health*, 45
(2019), pp. 233–5: https://blogs.bmj.com/bmjsrh/2019/10/16/
climate-change-and-contraception/

269   *'the conspicuous silence in recent years about the role of population growth'*
Aalok Ranjan Chaurasia, 'Population Effects of Increase in World
Energy Use and CO_2 Emissions 1990 to 2019', *The Journal of
Population and Sustainability*, Vol. 5, No. 1, 2020, 87–125.
https://jpopsus.org/full_articles_population-effects-of-increase-in-
world-energy-use-and-CO2-emissions-1990-2019/

269   *'providing more sustainable and energy-efficient refrigeration'*
Project Drawdown, 'Summary of Solutions by Overall Rank':
https://www.drawdown.org/solutions-summary-by-rank

270   *'TED talk given by Dr Katharine Wilkinson'*
Dr Katharine Wilkinson, 'How Empowering Women and
Girls Can Help Stop Global Warming', TED talk/video,
TEDWomen, November 2018: https://www.ted.com/talks/
katharine_wilkinson_how_empowering_women_and_girls_can_
help_stop_global_warming

272   *'as endorsed by a Special Report from the IPCC in August 2019'*
IPCC, 'Special Report on Climate Change and Land', August
2019: https://www.ipcc.ch/srccl/

272   *'4 Per 1000'*
https://www.4p1000.org/

272   *'the Terraton Initiative, has seriously raised the bar'*
The Terraton Initiative: https://terraton.indigoag.com/

273   *'the 54 leading countries are spending'*
As reported in Chuck Abbott, 'World Farm Subsidies Hit $2
Billion A Day', *Successful Farming*, 30 June 2020: https://www.
agriculture.com/news/business/world-farm-subsidies-hit-2-
billion-a-day

CHAPTER 18: TAKE HEART FROM HISTORY

280   *'The parallels are irresistible'*
Stephenson, *What We're Fighting for Now Is Each Other*, op. cit.

283   *'I do not wish to think, or to speak, or to write, with moderation.'*
William Lloyd Garrison, in the first issue of *The Liberator*, 1 January
1831; quoted in Sinclair, Upton (eds), *The Cry for Justice: An
Anthology of the Literature of Social Protest* (Philadelphia: The John C.
Winston Co., 1915); also available at Bartleby.com, 2010:
www.bartleby.com/71

284   *'Slavery is the holding of people at a workplace through force'*
Free the Slaves: https://www.freetheslaves.net/
our-model-for-freedom/faqs-glossary/

289   *'When you put your hand to the plow'*
Alice Paul, quoted in the 'Early Life' section of the

Alice Paul Institute website: https://www.alicepaul.org/
who-was-alice-paul/

CHAPTER 19: A JUST TRANSITION

290   *'Climate change is bigger than the New Deal'*
Roy Scranton, 'No Happy Ending: On Bill McKibben's "Falter"
and David Wallace-Wells's "The Uninhabitable Earth"', *Los
Angeles Review of Books*, 3 June 2019: https://lareviewofbooks.org/
article/no-happy-ending-on-bill-mckibbens-falter-and-david-
wallace-wellss-the-uninhabitable-earth/

292   *'We want to be front-runners in climate-friendly industries'*
Madeleine Cuff, quoting Ursula von der Leyen, '"Man on the
Moon Moment": Europe Unveils Plan to Be First Climate-Neutral
Continent by 2050', *BusinessGreen*, 11 December 2019.

293   *'climate change looming as the environmental justice issue of the twenty-
first century'*
Robert Bullard, quoted in Stephenson, *What We're Fighting for Now
Is Each Other*, op. cit.

293   *'those threatened by the cumulative stresses of climate change'*
'The Biden Plan to Build a Modern, Sustainable Infrastructure
and an Equitable Clean Energy Future', undated webpage: https://
joebiden.com/clean-energy/

295   *'a matter of basic fairness'*
Tom Kibasi, 'Taxing Income from Wealth the Same as Income
from Work Could Raise £90 Billion over Five Years, Report
Finds', press release, IPPR, 18 September 2019: https://www.ippr.
org/news-and-media/press-releases/slug-1306b49118532fe333771b
8f2c1b31d3

295   *'the treasuries and finance ministers of all OECD countries'*
Tim Jackson, *Prosperity Without Growth: Foundations for the Economy
of Tomorrow* (Routledge, 2017).

297   *'David Fleming was prescient in recognising the challenge this would pose'*
David Fleming's works include *Lean Logic: A Dictionary for the
Future and How to Survive It*, and a condensed version, *Surviving
the Future: Culture, Carnival and Capital in the Aftermath of the
Market Economy*, both edited by Shaun Chamberlain (Chelsea
Green Publishing Co, 2016). For more information about David
Fleming's work, see: https://www.flemingpolicycentre.org.uk/
lean-logic-surviving-the-future/

298   *'Men are qualified for civil liberty in exact proportion to their disposition'*
Edmund Burke, Letter to a Member of the National Assembly,
1791.

298   '"Getting Real about Immigration", inviting people who position
       themselves'
       Jonathon Porritt and Colin Hines, 'The Progressive
       Case for Taking Control of EU Immigration – and
       Avoiding Brexit in the Process', November 2017; http://
       www.jonathonporritt.com/sites/default/files/users/
       TheProgressiveCaseforTakingControlofImmigration.pdf

299   'I wrote five years ago that the first casualty of this African problem'
       Jeremy Grantham, 'The Race of our Lives Revisited', GMO White
       Paper, 2018, p. 25: https://www.divestinvest.org/wp-content/
       uploads/2018/10/GMO.the-race-of-our-lives-revisited.pdf

300   '2 million people a week already need humanitarian aid'
       International Federation of Red Cross and Red Crescent Societies,
       'The Cost of Doing Nothing: The Humanitarian Price of Climate
       Change and How it Can Be Avoided', IFRC Climate Center,
       Geneva, 2019; downloadable from: https://media.ifrc.org/ifrc/
       the-cost-of-doing-nothing/

301   'at least 212 environmental rights defenders were killed in 2019'
       'Defending Tomorrow', 29 July 2020: https://globalwitness.org/
       en/campaigns/environmental-activists/defending-tomorrow'
       2019: https://www.globalwitness.org/en/press-releases/
       spotlight-criminalisation-land-and-environmental-defenders/

301   'The true number is likely to be much higher'
       Patrick Greenfield and Jonathan Watts, 'Record 212 Land and
       Environmental Activists Killed Last Year', Guardian, 29 July 2020.

302   'there are now more than 3,000 conflicts over land and resources worldwide'
       Environmental Justice Atlas, accessible at: http://www.envjustice.
       org/ejatlas/

302   'It's a timely reminder of how the most effective solutions'
       Donna Lu, 'Guarding the Guardians', New Scientist, 13 March
       2021.

303   'Even now, WWF International is embroiled in a number of controversies'
       Tom Warren and Katie J. M. Baker, 'WWF's Secret War' series of
       articles, BuzzFeed, 2019–ongoing: https://www.buzzfeednews.
       com/collection/wwfsecretwar

303   'national recognition of land tenure, access and resource rights'
       Intergovernmental Science-Policy Platform on Biodiversity
       and Ecosystem Services (IPBES), media release, IPBES
       Secretariat, 7 May 2019: https://ipbes.net/news/
       Media-Release-Global-Assessment

304   'the crime of "ecocide" under the Rome Statute of the International
       Criminal Court'
       https://ecocidelaw.com/the-law/rome-statute/

304, 305   *'This is defined as: "the extensive loss or damage or destruction of ecosystem(s)"'*
      'Serious harm to the Earth is preventable.'
      Ecological Defence Integrity ('Stop Ecocide'): https://www.stopecocide.earth/; https://ecocidelaw.com/the-law/factsheet/

304   *'an official amendment to the Rome Statute that would criminalise acts'*
      Isabella Kaminski, 'Vulnerable Nations Call for Ecocide to Be Recognized As an International Crime', Climate Liability News, 6 December 2019: https://www.climateliabilitynews.org/2019/12/06/ecocide-international-criminal-court-vanuatu/

304   *'the EU and its Member States to promote the recognition of ecocide'*
      'EU Supports Recognition of Ecocide', Stop Ecocide, newsletter, 18 February 2021: https://www.stopecocide.earth/newsletter-summary/newsletter-february-2021-eu-supports-recognition-of-ecocide

CHAPTER 20: MOMENTS OF TRUTH

306   *'The all-too-real possibility that the story we're living is a tragedy that ends'*
      Scranton, 'No Happy Ending', op. cit.

307   *'To come / into knowing / is to come / into sorrow.'*
      Susie Orbach, 'Climate Sorrow', in Extinction Rebellion, *This Is Not a Drill*, op. cit., pp. 66–7, interpreted as a poem by Anthony Wilson, 'Climate Sorrow: A Found Poem, by Susie Orbach', 10 October 2019: https://anthonywilsonpoetry.com/2019/10/10/climate-sorrow-a-found-poem-by-susie-orbach/

309   *'Give up the fight with evolution. It wins.'*
      Catherine Ingram, 'Facing Extinction', blog, undated: https://www.catherineingram.com/facingextinction/

310   *'In August 2018, Observer journalist Will Hutton wrote an article'*
      As reported in Hopkins, *From What Is to What If*, op. cit.

311   *'As these messages are endlessly repeated and normalised'*
      Read and Alexander, *This Civilisation is Finished*, op. cit.

312   *'Jared Diamond's Collapse (published in 2004) catalogues a portfolio'*
      Jared Diamond, *Collapse: How Societies Choose to Fail or Succeed* (Penguin Random House, 2005).

312   *'Civilisations have throughout history marched blindly towards disaster'*
      Roy Scranton, 'Learning How to Die in the Anthropocene', *New York Times*, 10 November 2013: https://opinionator.blogs.nytimes.com/2013/11/10/learning-how-to-die-in-the-anthropocene/

314   *'Living our vocation to be protectors of God's handiwork'*
      Pope Francis, 'Encyclical Letter *Laudato Si*' of the Holy Father Francis: On Care For Our Common Home', *Vatican Press*, 24

May 2015: http://w2.vatican.va/content/francesco/en/encyclicals/documents/papa-francesco_20150524_enciclica-laudato-si.html

315   *'ecological sin against our common home'*
Pope Francis, 'Address of His Holiness Pope Francis to Participants at the World Congress of the International Association of Penal Law', 15 November 2019: http://www.vatican.va/content/francesco/en/speeches/2019/november/documents/papa-francesco_20191115_diritto-penale.html

316   *'We affirm that'*
'Islamic Declaration on Global Climate Change', The Islamic Foundation for Ecology and Environmental Sciences, 2.3, August 2015; downloadable from: http://www.ifees.org.uk/; PDF: http://www.ifees.org.uk/wp-content/uploads/2016/10/climate_declarationmMWB.pdf

317   *'Faiths interpret the world through story, poetry and art'*
Mary Colwell, 'Faith in Nature', *BBC Wildlife Magazine*, March 2016.

320   *'There is a wild and creative energy that flows through the universe'*
Mary Reynolds Thompson, *Reclaiming the Wild Soul: How Earth's Landscapes Restore Us to Wholeness* (White Cloud Press, 2014).

320   *'Some of the earliest academic work on this was done in the UK by The Conservation Volunteers'*
https://www.tcv.org.uk/greengym

320   *'researchers at Leeds Beckett University analysed the social value'*
Anne-Marie Bagnall, Charlotte Freeman, Kris Southby and Eric Brymer, 'Social Return on Investment: Analysis of the Health and Wellbeing Impacts of Wildlife Trust Programmes', Leeds Beckett University and the Wildlife Trusts, October 2019; downloadable from: https://www.wildlifetrusts.org/news/new-report-reveals-prescribing-nature; PDF: https://www.wildlifetrusts.org/sites/default/files/2019-09/SROI%20Report%20FINAL%20-%20DIGITAL.pdf

321   *'Reducing that deficit – healing the broken bond'*
Richard Louv, *Last Child in the Woods: Saving our Children from Nature Deficit Disorder* (Algonquin Books, 2005).

322   *'the seven underlying principles of harmony in nature'*
The Harmony Project: https://www.theharmonyproject.org.uk/

323   *'a love strong enough to break through the terror'*
Mary Annaïse Heglar, 'But the Greatest of These is Love', blog, Medium, 17 July 2019: https://medium.com/@maryheglar/but-the-greatest-of-these-is-love-4b7aad06e18c

323   *'Professor Ronald Purser refers to a kind of "quietist surrender"'*
Ronald Purser, 'The Mindfulness Conspiracy', *The Guardian*,

14 June 2019: https://www.theguardian.com/lifeandstyle/2019/jun/14/the-mindfulness-conspiracy-capitalist-spirituality; Ronald Purser, *McMindfulness: How Mindfulness Became the New Capitalist Spirituality* (Repeater Books, 2019).

CHAPTER 21: HOPE IN DISOBEDIENCE

326   *'Greater collaboration between governments and businesses is essential'*
      US Chamber of Commerce, 'Our Approach to Climate Change', webpage, undated: https://www.uschamber.com/climate-change-position

326   *'After a rain, mushrooms appear on the surface of the Earth'*
      Solnit, *Hope in the Dark*, op. cit.

328   *'Is it too late? We know what the science says'*
      Stephenson, *What We're Fighting for Now Is Each Other*, op. cit.

331   *'There is a far more pernicious form of science denial than Trump's'*
      Jonathan Safran Foer, *We Are the Weather* (Penguin, 2019).

333   *'Fossil fuel interests, and those doing their bidding'*
      Mann, *The New Climate War*, op. cit.

336   *'Either we have hope within us or we don't'*
      Václav Havel, *Disturbing the Peace* (Vintage, 1991).

INDEX